Saab Aircraft
since 1937

The Saab 37 shows its distinctive shape. *(Saab)*

Saab Aircraft
since 1937

Hans G Andersson

Smithsonian Institution Press
Washington, D.C.

© Hans Andersson 1989

Drawings by C G Ahremark

ISBN 0-87474-314-1

Printed and bound in Great Britain by
The University Printing House, Oxford

First published in the United States 1989
by Smithsonian Institution Press.

Library of Congress Catalog Number 88-064148

Contents

Editor's Note	6
Introduction	7
The Swedish Aircraft Industry – an Historical Survey	9
Saab 17	59
Saab 18	66
Saab 21	75
Saab 21R	81
Saab 90 Scandia	86
Saab 91 Safir (Sapphire)	93
Saab 29	104
Saab 32 Lansen (The Lance)	116
Saab 35 Draken (The Dragon)	125
Saab 105	137
Saab 37 Viggen (The Thunderbolt)	143
Saab MFI-15/17 Safari/Supporter	163
Saab 340	168
Saab 39 Gripen (The Griffin)	177
Appendix: FFVS J 22	182
Index	187

Editor's Note

In the Introduction and Historical Survey the reader will find the terms *Aktiebolag* and *Aktiebolaget*. *Aktiebolag* is Swedish for joint-stock company and the suffix *et* is the definite article. These words are frequently abbreviated to AB which may be taken as representing Ltd.

It will also be noted throughout this work that the definite article has been omitted before some aircraft type names. For example the Saab 32 is always referred to in Sweden as *Lansen*, which means the lance, the word for lance being *Lans*. The article in this instance is *en* – *en lans* equals a lance. The name and the article are not separated in referring to the type. This system also applies to *Draken*, *Viggen* and *Gripen*.

Introduction

This work describes the development of the Swedish aircraft and related industries. Despite the title, it actually covers more than 70 years of aeronautical development in Sweden.

Since Saab started in 1937 the company has produced more than 4,000 aircraft of thirteen different types. Of the Saab-designed types nearly 50 versions have been developed.

When Saab was established it was registered as Svenska Aeroplan Aktiebolaget (SAAB). In 1965, however, the company's name was changed to SAAB Aktiebolag to reflect the growing diversification of activities. Aircraft production no longer dominated, following the dramatic growth of motor car production. At the same time it was also decided that the name Saab could be used officially to identify the company, and this form has been used throughout the book.

This work has greatly benefited from the assistance of C. G. Ahremark who has provided the excellent three-view drawings of each type of Saab aircraft. I should also like to express my sincere thanks to Saab Aircraft, to the Swedish Air Force Staff's Information Department and to the individual Air Force Wings for providing many photographs.

Hans G. Andersson
Linköping, December 1988

Sweden's first aircraft industrialist, Dr Enoch Thulin, in appropriate setting in 1915. The aeroplane is his own Thulin D monoplane. *(A. Blomgren)*

The Swedish Aircraft Industry - an Historical Survey

Not surprisingly the history of the Swedish aircraft industry in general, and that of Saab in particular, is closely linked to that of Sweden's military aviation. Although in 1987 Saab celebrated its 50th anniversary as an aircraft manufacturer, it is necessary to go back at least another 20 years to form a proper perspective of Swedish aircraft industrial development and, not least, into government decision-making.

Aircraft manufacturing on an industrial scale - although modest - began in Sweden in 1914 at AB Södertelge Werkstäders Aviatikavdelning (SW) south of Stockholm. But two competitors, Svenska Aeroplanfabriken (SAF) in Stockholm and Aeroplanvarvet Skåne (AVIS) were not far behind. While SW mainly manufactured French Farman biplanes under licence and later the German Albatros trainer for the Army and Navy, and SAF manufactured the Albatros as well, AVIS had acquired a licence to produced the French Blériot monoplane (known as the Thulin A).

SW manufactured some thirty aircraft in the period 1914-17, including twelve each of the French and German designs, the latter, incidentally, having been copied from a confiscated example remaining in Sweden at the outbreak of the war. The Albatros became very popular and eventually as many as 54 examples and derivatives were built. A great disadvantage, however, was that Sweden was dependent on the import of (sometimes used) Mercedes engines from Germany, and it was only in 1916 that Scania-Vabis was able to start manufacturing six-cylinder Mercedes engines of 100/120 hp.

The SAF company was taken over by SW in late 1916 and at the same time yet another company, Nordiska Aviatik AB (NAB), was formed at Stockholm. In addition, NAB manufactured the Albatros for Sweden as well as for export. However, both SW and NAB failed to develop new competitive designs and therefore SW ceased operations in November 1917 and NAB seven months later.

A much more spectacular industrial venture was started in 1913 by Dr Enoch Thulin, initially under the company name AVIS. Thulin predicted a major expansion of both military and civil aviation and in 1914 formed a new company, AB Enoch Thulins Aeroplanfabrik (AETA), with considerable resources and the Swedish industrialist Gustaf Dahlén (inventor of the AGA light valve, etc) as the main sponsor. AETA was located at Landskrona in the south of the country.

In contrast to his competitors, Thulin was early aware that it was necessary for him to be able to supply the engines for his aeroplanes. During his pilot training in France, Thulin had had the foresight to buy a Le Rhône engine and was negotiating for a licence to produce the type when the war started. With able assistance from the Sandvik steel works, Thulin and his team managed to copy the engine and even improve its reliability, and its power was also gradually increased from 90 to 135 hp. As a result engine production actually became the backbone of the Thulin company.

Through Swedish Government procurement but also by building up a considerable export business with neutral nations, the Thulin factory had in 1918 developed into a major industrial complex with extensive design, test and production facilities for both aircraft and engines. By the end of 1918, the factory had produced nearly 100 aircraft of eleven different types, seven of which were of the company's own design. Some 650 engines were produced, mostly for export, with the Netherlands a major customer. The

An artist's impression of the Thulin works at Landskrona in 1918. The small inset drawing shows the factory in 1915. *(Landskrona Museum)*

Thulin company employed about 1,000 people and was, in fact, the largest aircraft/engine manufacturer among the neutral nations during the First World War. The Thulin engine team also designed and manufactured prototypes of a water-cooled 160 hp engine and in 1919 of a similar 260 hp engine, both of which can be seen at the Stockholm Technical Museum.

The Thulin designers, led by Dr Ivar Malmer, developed several notable designs of aircraft, some of which were even built under licence abroad. The company's last fighter design, the sleek Thulin NA biplane, powered by the Thulin/Le Rhône rotary engine, reached a top speed of 215 km/h (134 mph), but was too late for production. It can still be seen at the Landskrona Museum.

On 14 May, 1919, Dr Enoch Thulin unfortunately lost his life in a flying accident. This was a major blow to Swedish aviation, for without the 'flying doctor' in the pilot's seat his company was unable to survive as an aircraft manufacturer, especially since the government abruptly ceased ordering from private industry in 1919. Export contracts were also cancelled.

The Swedish Air Force started using the Dutch Fokker C.VE (S 6 in Sweden where it was also built under licence) reconnaissance biplane in 1927. The photograph shows an S 6 still airworthy in 1962. *(Saab)*

The Transition Period

When the Swedish Government in 1919 cancelled its procurement from the private aircraft industry, which eventually collapsed, Swedish military aviation had to support itself technically. Some of the experience gained during the Thulin era could, however, be saved by the Army's maintenance works at Malmslätt, Flygkompaniets Verkstäder Malmen (FVM), near Linköping. Since maintenance and repair work could not provide an even workload, aircraft design and manufacture had

Total production of Thulin engines (improved versions of the French Le Rhône rotary engine) was about 650, mostly for export. *(Landskrona Museum)*

already begun in 1918. Aircraft were also built under foreign licence. In the period 1918 to 1926 a total of 111 aircraft were manufactured at Malmslätt including 53 of Swedish design. Among the more numerous types were the Tummelisa (Ö 1) single-seat trainer, first produced in 1919 (30 built); the S 18 reconnaissance aircraft from 1919 (15); the J 1 Phoenix 122 fighter (25 built in Sweden, 15 imported) also from 1919; the Phoenix 222 (Ö 4/A1) reconnaissance aircraft and light bomber (14); and the S 21 reconnaissance aircraft (11) from 1925.

For the casual observer of the Swedish aircraft industry it may perhaps be easy to overlook the importance of these maintenance facilities for later, more spectacular, developments in the country.

The facts are that through a number of devoted personalities, the traditions were carried on from the Thulin years despite extremely meagre financial resources. Gösta von Porat and Peter Koch were able to establish the Malmslätt works without too much bureaucracy and the same applied to the aircraft designs created by Henry Kjellson and Ivar Malmer. The latter eventually became Sweden's first Professor of Aeronautical Engineering in 1928. Much later, in 1940, he became the head of Sweden's Aeronautical Research Institute (FFA).

Solving an Engine Problem

From 1919 the aircraft engine market was virtually flooded with cheap war-surplus engines, mainly from Germany. Although they enabled the Army and Naval aviation to acquire between them almost 200 aircraft and thus create a reasonable basis for the coming build-up of an independent air force, the highly varying quality and status of the surplus engines eventually became a serious flight-safety problem. Therefore, in 1923 the Government instructed the Army and the Navy to look into the matter of licence manufacture of modern engines. It took many technical and bureaucratic manoeuvres, however, before a licence agreement with the Bristol Aeroplane Company could be approved by the Swedish Government. Of the five companies competing for the production of the Bristol Mercury, Trollhättan-based Nydqvist & Holm (Nohab) – a subsidiary of Bofors, the armaments group – was selected. Nohab was awarded an initial contract for forty engines in April 1930, along with guarantees for eventual procurement of 300 engines over a 10-year period. For the purpose a special subsidiary company, Nohab Flygmotorfabriker AB, was formed. The first Mercury VI of 600 hp was delivered in 1933. No fewer than six

Forty Heinkel He 5 S and T (S 5 in Sweden) naval monoplanes were built under licence by Svenska Aero and Swedish military workshops starting in 1927. This picture shows an He 5T (S 5D) off Helsingborg ten years later. *(Flygvapnet/F 2)*

versions of the Mercury developing 600–980 hp were eventually delivered.

Foreign Birds

No record of the Swedish aircraft industry of the 1920s would be complete without mentioning Svenska Aero AB formed in 1921 at Lidingö near Stockholm and AB Flygindustri (AFI) formed in 1925 at Limhamn near Malmö in the south. The two companies were, however, not truly Swedish but in effect were subsidiaries of the German Heinkel and Junkers companies respectively, intended to preserve the capabilities of the German aircraft industry during the Allied restrictions imposed in 1918. Similar arrangements occurred in the USSR.

Svenska Aero was founded by the German naval pilot and engineer Carl Clemens Bücker (later famous for his successful long line of Bücker trainers from the 1930s which, incidentally, all had a Swedish chief designer, A. J. Andersson, who returned from Germany to Sweden and Saab in 1939). Svenska Aero manufactured the Hansa-Brandenburg naval reconnaissance sea monoplanes in several versions with Maybach and Rolls-Royce Eagle engines for the Swedish Navy. From 1927, a new design, the Heinkel He 5, powered by a Bristol Jupiter (later Mercury) engine was produced for the new-born Swedish Air Force (Flygvapnet). At the same time, licence-manufacture of the He 5 (Air Force designation S 5) by Flygvapnet's own work-shops began. Nearly 50 Hansa aircraft, as they were called in Sweden, were produced.

Svenska Aero, however, also established its own design office developing several aircraft types to Swedish specifications. With the Swedish engineer Sven Blomberg as chief designer, the company in 1929 completed prototypes of a fighter biplane, Jaktfalken, powered by a 425 hp Armstrong Siddeley Jaguar engine. The Swedish Air Force used one aeroplane under the J 5 designation; the other went to Norway. Later, the aircraft was re-engined with the Bristol Jupiter VI and VII as the J 6A and J 6B respectively. Svenska Aero also manufactured several trainer aircraft both of Swedish and Heinkel design. A biplane fighter on floats, the

Air Force Independence

The experience of the First World War called for air power, and in countries such as Great Britain and Italy independent air forces had already been established. This experience, notably from Great Britain, was extensively used in the Swedish defence debate in the early 1920s, notably by Carl Florman, a most eloquent supporter of air force independence. Carl Florman, incidentally, with his brother Adrian, also started ABA Swedish Air Lines, a predecessor of Scandinavian Airlines System, SAS, in 1924.

Svenska Aero's Jaktfalk (later J 6) fighter biplane, with Bristol Jupiter engine, made its first flight in 1929. The Swedish Air Force eventually ordered seventeen. The picture shows a J 6B delivered by ASJA in 1935. *(ASJA)*

One Jaktfalk was exported to Norway, fitted with an Armstrong Siddeley Jaguar engine and wheel spats. *(ASJA)*

Heinkel HD 19 (J 4 in Sweden) was delivered to the Swedish Air Force in 1928 (six aircraft). Altogether Svenska Aero manufactured 40 aircraft, 32 of them for Sweden.

In 1932 the company ran into financial difficulties and was taken over by AB Svenska Järnvägsverkstäderna (ASJA) of Linköping, C. C. Bücker returning to Germany. ASJA was mainly interested in the design team formed by Svenska Aero.

The Junkers subsidiary, AB Flygindustri, was formed with the sole purpose of producing Junkers aircraft for Sweden and for export. The Junkers technology was very advanced, using all-metal aluminium design, a novelty for Sweden. In 1924 ABA* (Swedish Air Lines) had started to equip with Junkers aircraft, beginning with the single-engine F 13 seaplane seating four passengers. Later on, larger three-engined aircraft carrying nine and later 16 passengers were produced by Flygindustri for ABA, these were the G 24 and the Ju 52/3m.

Flygindustri also produced military aircraft for export. These included the three-engined K 24 (1927) and the twin-engined K 37 (1928) bombers. The latter can be described as a predecessor of the later, better known Ju 86. In 1929, the company demonstrated a new two-seat fighter, the K 47, with a top speed of 290 km/h (180 mph). All these military aircraft were, however, designed in Germany and only assembled in Sweden.

After producing about 55 civil and 100 military aircraft (the latter all for export), Flygindustri ceased operations in 1935. Financially, it was not successful and the number of employees, which had been approximately 400 at the end of 1925, had dropped to less than 200 in 1935.

The 1924 Parliament (Riksdagen) was more positive in its approach to air power than its predecessors and in 1925 the then Defence Minister Per Albin Hansson – later Prime Minister – accepted the proposal for an independent air force, Flygvapnet. The Parliament approved the Government proposal in June 1925. A new era had begun. In brief, the decision was taken to provide the Air Force with a commander-in-chief, an air staff and an air board for technical/economic matters. The Service should comprise four combat Wings and one flying school. The Air Force should become effective on 1 July, 1926, and be fully established within five years. The maintenance and repair facility at Malmslätt (CVM) should become one of two central workshops, the second to be established at Västerås (CVV) in 1927. Both workshops continued to manufacture aircraft but on a limited scale. Between 1 July, 1926, and 30 June, 1936, a total of 109 aircraft were built,

*AB Aerotransport

42 of Swedish design and 67 under foreign licences. Of the latter, the Hansa (S 5) seaplanes and the Fokker C.VE (S 6) reconnaissance biplanes pre-dominated.

The original 1925 decision called for a strength of 229 aircraft to be procured over the coming 10 years. However owing to very meagre budgets, possibly in combination with the lack of maturity and tradition in the new Air Force command which delayed equipment matters, the actual aircraft strength in July 1936 was less than half of the original target. In 1925 the Government had accepted the necessity for a privately-owned Swedish aircraft industry. But it did not commit itself to any details or cost figures.

Following the 1925 defence decision several major Swedish industrial companies approached the Government expressing their interest in the manufacture of aircraft.

ASJA in the Lead

Although the German-owned Flygindustri was very active (but in the end unsuccessful) in securing a Swedish market also for its military aircraft, the most eager of the Swedish companies was AB Svenska Järnvägsverkstäderna (ASJ), in the main a manufacturer of railway equipment at Linköping, some 200 km (125 miles) south of Stockholm. Its managing director, Erland Uggla, had already been in contact with the Army in May 1924, but ASJ was also among the companies approaching the Government in February 1925. At a meeting with the Commander of Army Aviation, General K. A. B. Amundson, in September 1925, however, no firm procurement plans were presented. Consequently, the industry's initial significant interest diminished.

Only in September 1930 did the matter of aircraft manufacture again become a subject at an ASJ board meeting. Now, Uggla had 'come to the conclusion that aircraft manufacture could be started at little risk'. The Air Force, however, wanted a separate subsidiary company to be formed, and the board decided that AB Svenska Järnvägsverkstädernas Aeroplanav-

Elis Nordquist *(left)* **general manager of ASJA, and Lennart Segerqvist, chief test pilot, with a line-up of Hawker Hart (B 4) light bombers in 1936.** *(ASJA)*

delning (ASJA) should be established, with Sven Blomberg as general manager. Blomberg had been hired from Svenska Aero which was acquired by ASJ in 1932, but he left ASJA in 1934 to become managing director of Aeromateriel in Stockholm, a major import agency for manufacturers of foreign aircraft and equipment. At ASJA, he was succeeded by Elis Nordquist, Air Force Major (Engineering).

Although ASJA's main interest was in the military market, the first aircraft designed by the company was a single-engine three-seat cabin touring aircraft, the Viking I, which made its first flight in June 1931 powered by a 105 hp Cirrus-Hermes inline engine. Later on, it was re-engined with a 150 hp Walter Gemma radial and the second prototype was sold to a private company.

In 1934, a four-seat cabin touring aircraft, Viking II, powered by a six-cylinder de Havilland Gipsy Six en-

gine, made its first flight. A clean-looking aircraft, the Viking II prototype was used for a number of years by a Stockholm daily newspaper. Like its forerunner, it was flown with both wheels and floats. No production took place.

In the military field, ASJA started by designing and building to an Air Force specification two trainer biplane prototypes designated Ö 9. They were, in fact, seriously under-powered for the many requirements the aircraft was supposed to meet. ASJA also received a contract for delivery of twenty-five primary trainers of the German Raab-Katzenstein RK 26 Tigerschwalbe type. By a re-engining requested by the customer and resultant weight increases, the Sk 10 (the Air Force designation) suffered from rather sluggish spin recovery characteristics, causing several fatal accidents and leading to much painful public debate.

Contrary to ASJA's expectations, Flygvapnet's acquisition plans did not materialize at the pace predicted in 1930. In conjunction with ASJA's takeover of Svenska Aero in 1932 it received a contract for seven J 6B (Jaktfalken) fighters. Later, it also got an order for twenty-three Sk 11 trainers (D.H. Tiger Moths produced under licence) and, in addition, a promise of an order for eighteen B 4 (Hawker Hart) light bombers. But some time elapsed before these materialized.

At a board meeting in October 1934 ASJA's new managing director, Ragnar Wahrgren, presented his company's financial report, which the Board found very disappointing. In fact, the Board was willing to continue aircraft manufacture only if the Air Force could offer new information and assurances. However, another meeting which took place a month later with Defence Minister Ivar Vennerström and the Air Force C-in-C, General Torsten Friis, at which the industry complained about the few, irregular and uneconomical orders, would seem to have convinced the ASJA management that there was, after all, a future for Swedish aircraft production. Indeed, ASJA now decided to augment its resources and increase competitiveness for future contracts. Flygvapnet and many politicians wanted competition between more than two Swedish companies as well as from foreign suppliers. This revealed a certain degree of unrealism regarding basic industrial facts of life since it is, of course, not possible to achieve rational and economic production if reasonable production quantities cannot be obtained.

Despite the very limited funds available to the Air Force, yet another competitor to ASJA – at least with precious engineering manpower – appeared on the scene in 1933. Just after ASJA had taken over Svenska Aero, the Austrian engineer and aerobatic pilot Edmund Sparmann began aircraft design and manufacture in Stockholm. He was with the Phoenix-Werke in Vienna when he came to Sweden in 1919, but became a Swedish citizen in 1926 and was engaged as a test pilot for the Malmslätt works. Sparmann was also an inventor and had received payment

TOP: ASJA's Viking I, a three-seat touring monoplane powered by a Cirrus-Hermes, first flew in 1931. *(ASJA)*

ABOVE: The second ASJA Viking I had a 150 hp Walter Gemma radial. *(ASJA)*

from the United States for patent infringements during the war. From his private funds he started a design bureau in Stockholm, staffed mainly by German designers and draughtsmen. At the end of 1935 he employed nearly a dozen engineers/draughtsmen and an equal number of workers. In the late autumn of 1935 ASJA lost two engineers to Sparmann including Bo Lundberg, an Air Force engineering officer/pilot.

Sparmann's first venture was a light 'fighter trainer' which was completed in 1934. The Swedish Air Force did not particularly favour this concept but Sparmann somehow managed to convince the Government which in turn persuaded the Air Force to order four examples, three of which were paid for by Government funds allocated to combat unemployment. But Sparmann must also have had some supporters in the Air Force since later he won an Air Force order for a projected full-scale fighter, the E-4, an all-metal design. The Air Force considered Sparmann a skilled designer and wished him to remain in the Swedish aircraft industry; but he refused to collaborate with other companies and his own company was formally bought by Saab at Trollhättan in 1937 at the request of the Government.

The 1936 Decision

By 1935 it had become evident that a major strengthening of Sweden's defence was vital. More industrial resources were needed and firm long-term procurement planning necessary. In 1936 the Swedish Parliament approved a new decision regarding the defence forces covering the fiscal years 1936–43. For the Air Force the decision meant that during that period it would be equipped with 257 combat aircraft

LEFT: The ASJA Viking II of 1934 was a four-seater with a de Havilland Gipsy Six. It was acquired by *Stockholms-Tidningen*. *(ASJA)*

BELOW: The first military aircraft designed by ASJA, the Ö 9 multipurpose biplane with 330 hp Wright Whirlwind R-975E. Only two were built, in 1932 *(C. G. Ahremark)*

BOTTOM: Beginning in 1933, ASJA produced a series of German Raab-Katzenstein RK 26 Tigerschwalbe primary trainers under licence. *(Flygvapnet/F 5)*

and 80 trainers. Parliament also concluded that this procurement and later replacement aircraft should be acquired successively and according to long-term plans. The Swedish industry should have the capability to develop its own designs. In this connection, Parliament confirmed its earlier opinion that aircraft production should be undertaken by private industry while the military workshops should concentrate on maintenance and repair. But during the mid-1930s private industry still had to 'compete' for new orders with the military facilities.

Owing to the very limited production capacity of the private aircraft industry, Parliament accepted that at least during the next few years aircraft would have to be imported. In September 1936 the Air Force submitted a seven-year equipment procurement plan to the Defence Ministry. According to this, the Service was to procure 80 medium bombers, 90 light bombers, 55 fighters, 40 army reconnaissance aircraft, 32 naval aircraft and 95 trainers (55 primary and 40 advanced). Only 97 combat aircraft (40 bombers, 55 fighters and two naval aircraft) would be imported whilst 200 combat aircraft and all trainers would be procured from the Swedish industry. On 16 October the Air Force programme and the cost estimates were approved by the Government.

Competition, competition...

Besides ASJA several other financially sound companies approached the Air Force for orders, including Bofors (armaments and steel), Götaverken (shipyard), the Johnson Group (conglomerate) and Kockums (shipyard).

Competition was, naturally, a good thing but in view of the limited funds available there was also a major risk of splitting the resources into too many parts. This was a major issue for both the Air Force and the Government. There were many severe requirements to be met such as financing of the infrastructure; development resources, workshops, aerodromes, etc. The Air Force was only too accustomed to under-financed manufacturers repeatedly harassing the Service and other government agencies in order to survive. The effort to keep the total costs down seemed to call for maximum consideration, and Bofors, for one, was a strong supporter of this idea. In March 1936 Bofors had acquired Nohab Flygmotorfabriker which made it possible to concentrate virtually all aircraft and related manufacture (aircraft, engines,

Twenty-three de Havilland 82 Tiger Moth primary trainers (Sk 11s in Sweden) were produced under licence by ASJA in 1935-37. *(ASJA)*

weapons) at Trollhättan where Bofors declared a willingness to invest in workshops, aerodynamic and engine laboratories and an aerodrome. According to Bofors' managing director Sven Wingquist his basic formula was a 'solid, efficient industry, in strong consolidation and with a limited profit'. Regarding pricing, the international market should provide a healthy regulator. Bofors, incidentally, was of the opinion that a viable export of military aircraft was possible, and a firm supporter of this idea was industrialist Axel Wenner-Gren (founder of Electrolux, etc) who had an important financial interest in Bofors.

The Air Force C-in-C General Torsten Friis was always strongly in favour of competition but did not deny that he was attracted by the ideas promoted by Bofors. On the other hand, he found it unthinkable from a moral and practical standpoint to by-pass ASJA which had been contracted by the Air Force and also encouraged to expand its production and other facilities.

The ASJA management was also in favour of concentration but was unable to offer both aircraft and engine production. But ASJA had the great advantage over the competition by already being established as an aircraft manufacturer with a trained labour force, a design department and an aerodrome. However modest in size, there existed a nucleus that could easily be expanded. True, most of ASJA's experience was of aircraft built of steel and wood, but the company had long studied and had begun to prepare for production of modern light metal designs. There were no doubts about the financial strength as Stockholm's Enskilda Bank (the Wallenberg group) stood firmly behind the company.

The Air Force was now in the enviable position of having two powerful competitors ready and willing to meet its requirements.

The next step was not unexpected; the Bofors group offered an 'incorporation' of ASJA. Neither was the negative answer unexpected. Nevertheless, under Government pressure, the two groups jointly submitted a preliminary plan for a 'consortium' with factories both in Linköping and Trollhättan.

In December 1936 representatives of the Air Force and the industrial companies interested – not only ASJA and Bofors – were called to a meeting with the Prime Minister Per Albin Hansson. He informed them of the Government's basic views; the production could well be divided between two companies but since the funding was limited it was impossible to undertake development work at more than one organization. *All* activities had to be controlled by a

This Hart (B 4A) produced by ASJA in 1936 was powered by a 675 hp Swedish-built Nohab Mercury VIIA engine. *(ASJA)*

A total of 43 Harts was built in Sweden. They were used to introduce dive-bombing in the Swedish Air Force. *(Flygvapnet)*

central management. After an initial phase of licence-manufacture of foreign aircraft, Swedish designs meeting the requirements of the Air Force should be developed. The industrial companies were guaranteed – for the period 1937–43 – orders for 130 combat aircraft and 35 trainers; but after 1 July, 1943, the guarantees would expire.

It was now up to the industry to find a satisfactory way to co-ordinate their activities. Bofors had certainly not given up the idea of a dominating position but ASJA (Enskilda Banken) demanded equal rights.

An air view of the ASJA facilities at Linköping in 1937 *(centre)*. **The other buildings belonged to ASJ, the parent company.**

After protracted negotiations a joint development and management company AB Förenade Flygverkstäder (AFF) was formed in Stockholm in March 1937 by ASJA and Saab. The latter company was then in the process of being established by the Bofors group at the direct request of the Government. The formation of AFF was not a happy marriage but at least some groundwork had been laid for the new Swedish aircraft industry.

With the prior approval of the Government, on 10 April, 1937, the Air Force Board signed a contract with AFF which then began its activities.

The top management of Saab at Trollhättan in 1938. *Left to right*: Claes Sparre, production manager; Axel Wenner-Gren, chairman of the board; and Gunnar Dellner, managing director. *(Saab)*

Saab Established

SAAB (Svenska Aeroplan Aktiebolaget AB) was incorporated at Trollhättan on 2 April, 1937, and four days later, the first shareholders' meeting was held. The share capital was set at 4 million Swedish Crowns of which 1.5 million belonged to Bofors-Nohab and 2.5 million to AB Ars which was part of the Electrolux group representing Axel Wenner-Gren. Wenner-Gren was elected chairman and Gunnar Dellner managing director. Saab also took over the Nohab shares in Nohab Flygmotorfabriker. A new factory for Saab was built immediately north of Trollhättan and at the same time new facilities were built next-door for Nohab Flygmotorfabriker. In 1938 the share capital was doubled. In the meantime new facilities were built by ASJA at Linköping and both were opened in 1938. Trollhättan was selected for the manufacture of medium bombers and Linköping for light bombers and trainers.

In November 1936, the Air Board had secured a manufacturing licence from Focke-Wulf in Germany for the Fw 44J Stieglitz (Sk 12) primary trainer. Apart from aircraft ordered directly from Germany, in 1937 the Air Board contracted ASJA for twenty Sk 12s. Since that contract had been prepared before AFF was formed, it was signed be-

The Focke-Wulf Fw 44J Stieglitz trainer was introduced in the Swedish Air Force in 1936. In 1937, ASJA received an order to produce it under licence as the Sk 12. A very popular aircraft, it was used by the Swedish Air Force until 1946. A total of 77 was acquired from various sources. *(Flygvapnet/F5)*

tween the Air Board and ASJA. Between July and September 1938 the following contracts were awarded to the industry via AFF:

40 Junkers Ju 86Ks (B 3s) by Saab, and the same number was ordered from Junkers
40 Northrop 8A-1s (B 5s) by ASJA
35 North American NA-16-4Ms (Sk 14s) by ASJA

The necessary licence agreements had been signed in 1937 with Junkers Flugzeug- und Motorenwerke AG, Nortrian engineer Alfred Gassner, previously with Junkers and Fokker, as their chief designer. He and his team were posted to Stockholm and the AFF design office there. But rivalry prevailed between the parents.

In the meantime, ASJA had further strengthened its lead over the Bofors group by hiring no fewer than 46 United States designers and stress specialists. They were recruited starting in May 1938 by James A. Burnett, an experienced engineer from the Douglas Aircraft Company, and by his successor, Ernest W. Kazmar. Three groups came in 1938–39 mainly from Douglas companies. The recruitment campaign was not greeted with too much enthusiasm by the US State Department as there was indeed a shortage of such specialists even at Douglas. For this reason restrictions to work abroad were later introduced for people working under US defence contracts.

The hiring of the Americans was understandable in view of the power play that led to the formation of AFF. ASJA mistrusted the Bofors group and wanted to safeguard itself against a possible Bofors 'coup'. This mistrust was strengthened by the fact that the AFF designers were not formally employed by AFF but by Saab at Trollhättan and loaned to AFF. The Americans' stay in Sweden, however, was to be brief as they were recalled when war started in Europe, the last man leaving Sweden in March 1940. They had, however, transferred much valuable design experience to their Swedish colleagues.

The friction between ASJA and Saab was further increased in connection with a design competition early in 1938 for a new Army and Naval reconnaissance aircraft. Good visibility was important for the observer and the aircraft could well be high-winged. The

In late 1938, ASJA received a Swedish Air Force order for an initial thirty-five North American NA-16-4M (Sk 14) advanced trainers (455 hp Wright Whirlwind). Eventually as many as 136 aircraft were produced under licence by ASJA and later Saab. The picture shows the Sk 14A built by Saab in 1942 powered by a 500 hp Piaggio P VII RC-16 engine. *(Saab)*

throp Corporation and North American Aviation Inc.

In order to undertake these contracts but above all to speed up expansion of the companies' design departments and familiarize them with light metal stressed-skin designs, both Saab and ASJA engaged foreign engineers.

The Air Force's intention that AFF should become the major project and design organization could never be fulfilled, however. AFF was owned fifty-fifty by ASJA and Saab and therefore had no power to control the ambitions of the two parent companies. To make the Air Force ambitions a reality, the ASJA development resources would have had to be transferred to AFF. But this ASJA refused. Saab hired the Austop speed had to be at least 400 km/h (250 mph). The AFF/Gassner project, as it was known, was externally somewhat similar to the British Westland Lysander and featured stub wings housing a retractable undercarriage or used for attaching floats. The wind-tunnel testing, however, showed that the stub wings led to aerodynamic disturbances and they had to be abandoned. The result was a high-wing aircraft with non-retractable undercarriage, virtually impossible to modify into a light bomber which was another requirement. ASJA, which would have produced the aircraft had it been accepted by the Air Force, also evaluated this project with a very negative result both regarding performance calculations and general design. The Gassner project was cancelled.

Meanwhile, ASJA had submitted to the same requirements a project known as the L-10 (later Saab 17). The basic idea here was to develop a multi-role aircraft that could be used for bombing and for Army and Naval reconnaissance. Perhaps the view for the observer was not ideal but the advantages of a standardised type for all three kinds of missions would be very great in production, training, technical support and – not least – economy. It was an advanced design, with retractable undercarriage fairings that could be used as aerodynamic brakes in dive bombing; retractable skis; integral float design; internal bomb stowage. At a mock-up inspection and design review, the project was approved by the Air Force and two prototypes were ordered on 29 November, 1938. A great day for the young design team at Linköping!

The 'New' Saab is Founded

Despite the existence of AFF the mistrust between ASJA and Saab showed that a collaboration between the two in the development field was almost impossible. Torsten Nothin, a former Member of Parliament and now chairman of the AFF board, realized that a radical new approach was urgently needed. At a board meeting in December 1938 he strongly urged the two parties to reconsider the whole issue. The two companies were given until January to submit their views. The eventual result was that in March 1939 Saab was completely restructured. The company was separated from Nohab and, in effect took over ASJA. Saab became an aircraft manufacturer only. Nohab Flygmotorfabriker concentrated on engines only. The new Saab company (or rather Svenska Aeroplan Aktiebolaget, Saab) was to have a share capital of 13 million Swedish Crowns. The new shares were all purchased by AB Svenska Järnvägsverkstäderna, ASJ, the parent company of ASJA. Torsten Nothin was elected board chairman of Saab, and also on the Board was Marcus Wallenberg Jr of Stockholms Enskilda Bank who had been the driving force behind ASJA all along. His role in establishing a Swedish aircraft industry cannot be overrated.

Ragnar Wahrgren became managing director of the new Saab company and Linköping the headquarters for the company. All design and development activities were concentrated there, the Trollhättan factory being retained for production only.

Regarding the development resources, there was initially considerable Air Force pressure on Saab to retain Alfred Gassner as chief designer, but Saab resented this strongly as the two existing engineering teams were really not compatible especially in view of the strong American technical influence at Linköping. Eventually, Gassner returned to Fokker. AFF had lost its importance.

When the Second World War began in September 1939, none of the aircraft ordered in 1938 had been delivered by the Swedish industry. In the autumn of 1939 and early in 1940 the possibility of importing combat aircraft quickly faded. Approximately 195 combat aircraft, many approaching obsolescence, was all that the Swedish Air Force could mobilize on 1 September, 1939. Aircraft on order in the United States, The Netherlands, France and Germany never reached Sweden. Only Italy remained open during 1940. Flygvapnet's situation could only be described as desperate.

More Aircraft Needed

In March 1939, the Defence Minister, Per Edvin Sköld, had written to Saab asking whether the company could increase its production capacity by 50 percent. Saab replied that the company was prepared to do so on three conditions: that
1) Swedish manufacturers should be given the opportunity to bid on all Air Force orders. Orders should be placed in Sweden as far as possible and foreign offers with 'dumping prices' should not be considered.
2) No monopoly was requested, but only bona fide offers from credible Swedish suppliers should be considered.
3) The company should be allowed to export its products in the same manner as the Swedish armament industry.

The bargaining position of Saab was now fairly strong. The Air Board and the company eventually negotiated a 10-year policy agreement which in fact gave the company, although not formally, a monopoly on aircraft development and production in Sweden. The agreement was approved by the Government on 6 October, 1939. A similar agreement was signed with the engine industry.

It may still be of interest to recall in this context that only in late 1939 did Götaverken (GV) finally give up its ambition to compete with Saab. The company had made a modest start in 1935 by building a few Hawker Hart (B 4) light bombers for the Air Force. In 1938, the Air Board invited the company to submit a proposal for a twin-engined bomber (in itself a rather surprising move in view of concentration efforts behind the AFF). In January 1939 Bo Lundberg had joined GV as chief designer, subsequently hiring as many as 16 engineers. The GV GP 8 bomber project was regarded by the Air Force as the most advanced but it also proved to be by far the most expensive of the three proposals submitted. Even if the unit price was later reduced, the Air Board demanded that a new factory be built by GV before a prototype order could be awarded. For the GV Board, however, the prospect of building a prototype was not enough to justify financially a new factory.

But there is more to be said about GV and Lundberg. In April 1939 the Air Board had also invited both Saab and GV to submit proposals for a new fighter aircraft to be powered by the Bristol Taurus sleeve-valve radial engine of 1,215 hp. The Board specified tenders by 15 September, 1939, first flight by 1 July, 1941 and first delivery by 1 July, 1943. Both proposals arrived on time. But at the same time the GV informed the customer that it had decided to withdraw from aircraft manufacture and that it had transferred its rights and responsibilities to a new company, AB Flygplanverken (AFV) formed by Lundberg.

Both projects, the Saab L-12 (later J 19) and the AFV GP 9 were of fairly

Another licence manufacture by ASJA/Saab was the Northrop 8A-5 (B 5 in Sweden) dive-bomber, a rugged aircraft, here fitted with skis. *(Flygvapnet)*

In 1940-41, Saab delivered a total of 102 B 5s. *(Flygvapnet/F 4)*

conventional design but the GP 9 top speed was, surprisingly, given as high as 682 km/h (424 mph) against 605 km/h (375 mph) for the L-12. The cost of the GP 9 was 30 percent higher. The Air Board declared itself willing to order prototypes of both aircraft but again on the condition that AFV could arrange suitable production premises. Despite intensive efforts by Lundberg this proved impossible. Soon thereafter Lundberg was called back into the Air Force. The Air Force eventually decided that the Saab L-12 was too much for Saab to handle on top of the two bomber projects and in December 1939 both fighter projects were shelved. Saab concentrated on the bombers already contracted for and the Air Force was – at least for the moment – convinced that the fighters could be imported from the United States.

A Tough Job

Rome was not built in a day and the same could certainly be said about the Swedish aircraft industry. The start of licence production at both Linköping and Trollhättan was – not surprisingly – rather slow. The build-up of experience among both engineers and factory workers, not to speak of those of the many suppliers, was a painstaking and time-consuming operation. Sweden had no traditions in the field and the problems were not always understood by the industry's sometimes harsh critics. One example: According to the original plans, forty B 5 (Northrop) light bombers were to have been delivered between August 1939 and the autumn of 1941. Initially, ASJA/Saab had planned to import some important components. Owing to the outbreak of the war, this became either difficult or outright impossible. The supply of drawings, etc, from the United States was delayed and all this taken together forced Saab and the Air Board to agree on a new delivery schedule. The start of the deliveries was now set for the end of 1939 but completion must be before 15 January, 1941. In the event, all 40 aircraft were delivered during 1940.

The B 3 (Junkers Ju 86K) was already obsolete in 1939 (one may now wonder why it was chosen for licence production in mid-1938) and therefore no efforts were made in 1939 to accelerate its production at Trollhättan. On the contrary, the Air Board was authorized by the Government in February 1940 to stop production of this slow bomber which would have been no match for modern fighters. Instead, the Air Board was authorized to negotiate for additional B 5 light bombers. Eventually, a total of 102 B 5s were delivered.

During the first months of the war Sweden's strategic position became extremely critical. On 30 November, 1939, the Soviet Union attacked Finland and invaded the three Baltic States. On 9 April, 1940, Germany invaded Denmark and Norway, Denmark being occupied within 12 hours, and Norway in eight weeks. From a Swedish point of view it was particularly serious that the Soviet Union and Germany now acted in concert following the famous non-aggression pact signed in August 1939. As a consequence, the Germans refused to deliver war material to Sweden as long as the Soviet Union was at war with Finland. Everyone knew that Sweden had assisted Finland with as much equipment as the country could possibly afford and even sent voluntary combat forces including a mixed fighter/light-bomber squadron with a total of 17 aircraft (12 Gladiators and five Harts).

Urgent Work For Finland

Like Sweden, Finland had ordered a considerable number of combat aircraft abroad in 1939 but few of these had arrived by the time of the Soviet attack. Finland urgently appealed for military assistance from the Western Powers. In mid-December the Finnish authorities requested Swedish assistance in providing a large number of aircraft, with assembly, equipment and transfer to suitable bases for later delivery to Finland. There proved to be so much work involved that a special section for Finnish matters had to be organized within the Air Board. The original planning called for as many as 460 aircraft.

The first aircraft for Finland, thirty Gloster Gladiators from Britain, arrived in January 1940. They were rapidly assembled by the Air Force works at Malmslätt and fitted with Swedish skis. Then followed a stream of aircraft that were assembled at Malmslätt, by Saab at Trollhättan, and by ABA Swedish Air Lines at Malmö. Saab

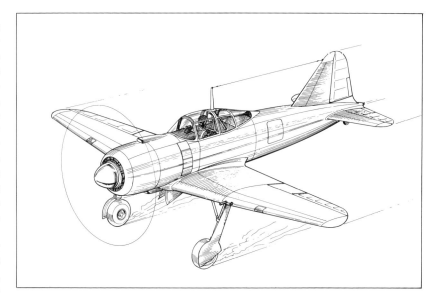

TOP: The Saab L-12 (later J 19) was a well-defined project for a single-engined fighter to be powered by a 1,215 hp Bristol Taurus engine. It had an estimated top speed of 605 km/h (376 mph) but work was suspended in December 1939. *(C. G. Ahremark)*

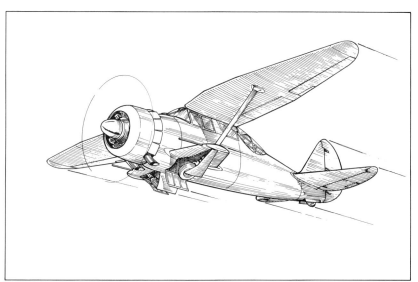

ABOVE: The AFF P7 project for a reconnaissance aircraft was defeated by the ASJA L-10 which became the Saab 17. *(C. G. Ahremark)*

reported to the Air Board a delay of about three months in the B 3 production because of the Finnish programme which for Saab included 44 Brewster Buffaloes and 17 Fiat G.50 fighters, the latter being assembled at the rate of one a day. The foreign aid originally promised dwindled, however, and in the end a total of 116 aircraft were assembled and equipped in Sweden.

An interesting development in the

The Junkers Ju 86K (B 3 in Sweden) twin-engined bomber was introduced in the Air Force in 1937. In mid-1938, Saab was contracted to build forty B 3s under licence at the new Trollhättan factory. *(Flygvapnet)*

Swedish Aid-Finland campaign was the public fund-raising for the purchase of fighters. After consultation with the Finnish Air Force the Swedish Air Board was able to secure twelve Fiat CR.42 fighters which according to the agreement would be delivered by 1 April. But now the war had ceased (at least temporarily) and the Finnish interest in the Italian biplane fighters diminished. Not suprisingly, the Air Board received a message from Helsinki that the Finnish Air Force would prefer to have the money. The CR.42s thus remained in Sweden and fifty additional aircraft were ordered mainly for Air Force training.

The Great American Hope

To acquire aircraft from abroad rapidly became almost impossible. Sweden tried very hard indeed in 1939-40 to order defence material in the United States since Britain and Germany refused in particular to sell modern fighters. Obviously, the United States needed to keep equipment for itself, and had also undertaken to give supply priority to Great Britain and France. It had been decided by the US Government that Great Britain and France

In early 1940 production of the B3 was stopped in favour of additional light bombers. In the late 1940s, the B3 was used to carry torpedoes. *(Flygvapnet)*

should be supported in the war against Germany, and Sweden's ability to withstand German pressure after the occupation of Denmark and Norway was cast in doubt by Washington. The Swedish declaration that the equipment was needed for maintaining strict neutrality against Germany was not accepted. On 2 July, 1940, Sweden's orders in the United States became the subject of an export prohibition and on 10 October the same year a law enforcing confiscation of war material came into effect.

By this time Sweden had placed orders in the United States for: 120

A few B 3s were converted into 10-passenger transports in the mid-1940s.

Seversky-Republic EP-1 (J9 in Sweden) fighters including a manufacturing licence; 52 Seversky-Republic 2 P-A (B 6 in Sweden) light bombers; 144 Vultee 48C fighters; 550 Pratt & Whitney Twin Wasp engines; plus ammunition, machine tools, etc. Of a total of 316 aircraft on firm order, only 62 (60 EP-1s and two 2 P-As) actually arrived in Sweden.

The negotiations for licence manufacture of the Twin Wasp and the Double Wasp were broken off. Sweden eventually managed to copy the Twin Wasp from DC-3 engines; but there was no Double Wasp to copy. The great hopes pinned on the United States as *the* supplier of fighter aircraft virtually disappeared overnight.

The only source for combat aircraft still open in 1940 was Italy where 132 fighters (72 Fiat CR.42s, 60 Reggiane Re 2000s) and 84 Caproni Ca.313 bombers could be obtained. The Fiats (J 11 in Sweden) were the last fighter biplanes and already obsolete. But without these and the Re 2000s (J 20), two fighter Wings (six squadrons) could not have been established. The Caproni bombers (B 16/S 16) contributed to a dark chapter in Swedish Air Force history. They suffered as many as sixteen fatal accidents in which 41 Swedish airmen lost their lives, although in fact not all the accidents were due to equipment failures. The Italian order, which also included nearly 200 Piaggio engines, *was* an emergency action and vitally necessary in 1940.

This fairly extensive description of the import situation in 1940 has to be taken into account when discussing the build-up of the Swedish industry and the urgency prevailing in 1939-40.

The Air Force Board was no longer in complete control of the situation. The appointment of a Supreme Commander (General O. Thörnell) with special powers in December 1939, and at the same time the creation of a Government Industrial Commission (IK) responsible not only for the supply of strategic materials but also for the functioning of the defence industry, affected the military decision-making.

The Air Force had strongly criticized Saab for its slow production build-up and the Industrial Commission immediately conducted a thorough study. The problems were numerous but above all the lack of trained personnel and materials dominated. Saab needed at least 600 engineers/draughtsmen to develop the two aircraft types already contracted for by the Air Force. The 1940 level was less than 400, and the 600 level was, in fact, not reached until early 1943.

The Industrial Commission realized that only by concentration could the problems be solved – not by competition as some headstrong Air Force officers still insisted.

In early 1940 the Air Force had submitted a new plan for yet another major expansion of the number of combat Wings, and, on 8 August, 1940, the Supreme Commander requested and received Government permission to establish five new Wings (15 Squadrons) in addition to the nine Wings already existing or in the process of being organized. The Air Force was now totally dependent on the Swedish industry, and its expansion took priority over almost everything else in the rearmament programme.

The First Basic Agreement

After several months of negotiations, a mutual Basic Agreement (in Swedish, *Ramavtal*) between the Government and the industry was submitted to the Government for approval in November 1940. This agreement, which had been worked out by the Industrial Commission (IK), laid a solid foundation for the coming expansion of the Swedish aircraft industry as a reliable supplier to the Air Force. Saab accepted development of the aircraft types specified by the Air Force and promised to start production deliveries 33 months after

In May 1940, the prototype of the Saab 17 light bomber/reconnaissance aircraft, made its first flight. The photograph, however, shows the last version produced of this successful design, the B 17A which was powered by a Swedish-built Pratt & Whitney Twin Wasp engine. *(Flygvapnet/F 6)*

receiving development contracts. The production rate could be set as high as 30 aircraft a month if required. The agreement covered deliveries of not fewer than 1,100 combat aircraft by mid-1946. Underground factories were also included in the agreement.

The Air Force Board, the aircraft/engine industry and the Industrial Commission soon developed excellent teamwork. The key personalities were Colonel (later Major General) Nils Söderberg as head of the Air Material Department (on the Air Force Board), banker Marcus Wallenberg Jr. (for the Saab board) and Uno Forsberg, an experienced industrialist especially appointed by the Industrial Commission to deal with aircraft/engine production.

Engines – the Critical Factor

For the aircraft designers, the lack of suitable engines proved to be a serious

Major expansion of the Swedish aircraft industry took place in 1940 and an agreement was signed with the Government calling for delivery of more than 1,000 aircraft over a few years. In this picture taken in 1942, some of the total of 322 Saab 17s are being assembled. *(Saab)*

bottleneck and compensation could not easily be made for lack of engine power, although great efforts were made. Licence-manufacture of Mercury engines had already begun in 1930 at Nohab under the Bristol Mercury agreement. Until 1936–37, however, the capacity was limited to 30 engines a year. In January 1937, the Air Board had requested a capacity increase to 70 a year but even in September 1939 the annual production was less than 60. The problems at Nohab, which actually led to a management shake-up, prompted the Air Board to approach motor car manufacturer Volvo in an effort to involve it in aircraft engine production. It took until 1941, however, before Volvo took over Wenner-Gren's majority shareholding and changed the company's name to Svenska Flygmotor AB (SFA).

Sweden's isolation during the Second World War hit the engine industry particularly hard and, contrary to earlier hopes, it proved to be impossible to develop in reasonable time, with the resources available, the engines required by the Air Force. In fact, starting in 1940 SFA managed to copy the Pratt & Whitney Twin Wasp, in itself a formidable task which the Americans did not consider possible. Despite such a technical challenge, Flygmotor achieved an annual production of 313 engines in 1944. Take-off power was 1,065 hp and reliability every bit as good as the original.

A 1940 dilemma was the failure of engine standardization hoped for in 1939 when the Twin Wasp and Double Wasp licences were still being negotiated for the three new combat aircraft under development at Saab. The Twin Wasp copy did not become available until 1943 and even then in very limited quantity. Therefore this engine had to be reserved for fighters; the early bomber/reconnaissance aircraft had to be powered by other engines, notably the licence-built Bristol Mercury XXIV of 980 hp and the imported Piaggio P XIbis RC 40 of 1,040 hp. The Twin Wasp could be made available only for the last version of the Saab 17.

The serious engine availability problems caused new delays in the delivery schedules agreed upon in the Basic Agreement from late 1940 and the schedules had to be revised downwards in the autumn of 1941. The IK had to act again ordering new priorities for both the aircraft and engine industry.

The Saab 17 (formerly L-10) dive-bomber and reconnaissance aircraft, which had been ordered into development in late 1938, made its first flight in May 1940. The original number of aircraft to be produced had been set at 260 in the 1940 Basic Agreement, with production deliveries starting in early 1942. The number of aircraft was later increased and in October 1944 the last of 325 was delivered. Four different versions with three different types of engines and three different types of undercarriage were produced.

Ambitious Planning

The Air Force and industry planning in 1938–39 was indeed very ambitious in view of the limited experience and resources available. As early as the beginning of 1939, a development contract was awarded to Saab for a twin-engined bomber, the Saab 18 (originally the L-11), also intended for strategic reconnaissance and torpedo-launching. At the outbreak of the war,

In early 1944, production of the twin-engined Saab 18 bomber was in full swing. *(Saab)*

Fly past by sixteen B 18As *(Flygvapnet/F5)*

The Saab 18 was an elegant design. The B 18A/S 18A was powered by two Svenska Flygmotor-built Pratt & Whitney Twin Wasps each of 1,065 hp. Note the offset position of the cockpit canopy introduced to improve the pilot's downward visibility. *(Saab)*

the development work was temporarily halted, partly to enable a concentration of resources on the Saab 17, partly to enable the designers to include some modifications based on early wartime experience. In June 1940, development work was resumed. The first prototype flew in June 1942, with production deliveries starting in March 1944. A total of 244 aircraft was produced.

Owing to the engine shortage, the Saab 18 had to be designed in two main versions. The first one, the B 18A/S 18, was powered by two SFA Twin Wasps, but a more powerful version – initially the Bristol Taurus was planned – could not fly until 1944, equipped with the Daimler-Benz DB 605B. This version, the B 18B, and the similar T 18B, were fast and had a top speed of 570 km/h (355 mph).

A Revolutionary Bombsight

The Swedish Air Force was one of the pioneers of dive-bombing, starting in 1934 using the Hawker Hart (B 4) and later the Northrop 8A-1 (B 5). The Air Force was also well aware of the limitations of dive-bombing, especially the bomber's vulnerability against anti-aircraft fire and fighters. Saab took a great interest in the problem and in the early 1940s two engineers, Erik Wilkenson and Torsten Faxén, solved the problem, which eventually resulted in a remarkable increase in the efficiency of the Swedish bomber/attack

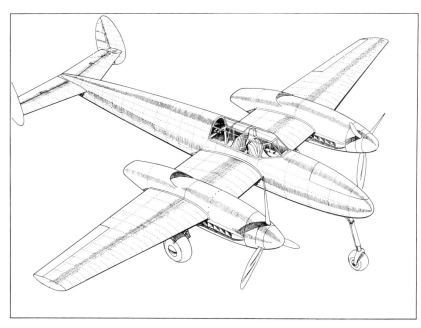

Saab 24 was the designation of a projected twin-engined (Daimler-Benz 605Bs) dive-bomber and night fighter with an estimated top speed of 630 km/h (392 mph) and a loaded weight of 7,895 kg (17,400 lb). A full-scale mock-up was completed but further development cancelled in December 1943. *(C. G. Ahremark)*

aircraft. In contrast to the steep – up to 90 degrees – dives used by the German Stukas, the new bombsight allowed shallow dive angles (20–30 degrees) and with automatic bomber separation during the pull-out. The bombsight, which was developed under strict secrecy, was a mechanical computer – one of the first of its kind in the world – using target-data input from the sighting phase. The sighting system permitted moderate altitudes to be used before the final diving phase thereby considerably reducing the time of exposure to anti-aircraft and fighter defence.

The first production 'toss' bombsight, the BT 2, went into the B 17. A later design, the BT 9, went into the B 18 and was also used in the A 21A. After the Second World War, new mixed mechanical/electronic versions were developed, sold to and even licence-produced in the United States. The BT 9 was also exported in sizeable numbers to France, Denmark and Switzerland. The last Swedish aircraft to carry the BT 9C was the Saab 32 Lansen first introduced in to service in 1955. Saab produced nearly 2,000 sights not including US licence production or 'pirate versions' produced there after the Swedish patents expired. The BT 9 represented Saab's first major equipment development programme outside the traditional aircraft design area. The development of the first Saab pilot-ejector seat used in a production-type aircraft as early as 1943 was another contemporary technical milestone.

The Urgent Need for Fighters

In line with the international doctrines on air warfare prevailing in Europe in the 1930s (the widespread influence of the Italian General Guilio Douhet and his deep and much publicized belief in the offensive elements), the 1936 parliamentary defence decision in Sweden called for very few fighters in relation to bombers. The ratio was, in fact, 4:1 in the bomber's favour. This emphasis on 'strategic' bombers and reconnaissance aircraft was also motivated by Sweden's geographical situation. The main threat was an invasion mainly over the long sea borders. Such an attack could not be stopped by fighters, it was claimed. The fighters were only considered important for local defence. Only in February 1940 had the Government authorized the establishment of two additional fighter Wings over the single one existing in early 1940.

Saab was deeply involved in the bomber programmes and could not be disrupted by fighter production, however urgently needed. The Industrial Commission (IK) was also firmly against it.

The Last Improvisation

In the late summer of 1940, Colonel Nils Söderberg, the head of the Air Board's Material Department, wrote to the resident Swedish engineer at Vultee, Bo Lundberg, who was attached to the Swedish Air Commission in the United States, and asked him to prepare for the design of a 'stop-gap' fighter aircraft. Owing to the shortage of aluminium which hampered Saab's work, it was decided that the aircraft should be built largely of steel and wood. Engineers should be recruited mainly from the Air Force and from companies not already working as suppliers to Saab, at least as far as possible. In addition, some work should be done by the Air Force's own workshops, which were still quite experienced in those materials. The Söderberg/Lundberg plans for a fighter designated the J 22 and based on the use of the Swedish version of the Pratt & Whitney Twin Wasp engine, was generally well received in the Air Force and by defence politicians, and initial funding for prototypes was authorized in February 1941.

Following the 10-year policy agreement signed in 1939 with the aircraft/engine industry, Saab had to be consulted. The company's initial reaction was somewhat negative. The managing director Rognar Wahrgren, in fact, told the Air Board that the company was not interested in the manufacture of an aircraft which in his opinion would not be 'first class' (owing to the limited engine power available; he even suggested that the Republic EP-1 be copied like its engine). Saab, however, assisted with weight and performance calculations. The estimated performance – a top speed of 575 km/h (357 mph) – seemed optimistic and, in addition, Lundberg's patented structural design, with load-carrying interchangeable wooden panels, was very advanced.

For the assembly work a new hangar was built at Stockholm's Bromma Airport for later use by ABA Swedish Air Lines, using government unemployment funds. The facility was leased by the Air Board from 1 July, 1942 to 1 July, 1945. The first J 22 prototype was completed on 1 September, 1942, for ground testing and on 20 September

Lieutenant General Bengt G. Nordenskiöld, dynamic Commander-in-Chief of the Swedish Air Force 1942-54 (LEFT) and Major General Nils Söderberg (RIGHT), head of the Swedish Air Force Board 1944-50, played a major role in the building-up of the efficient teamwork between the Air Force and the industry. *(Flygvapnet)*

The T 18B was the last version of the Saab 18 produced. With a top speed of 595 km/h (370 mph) it was one of the fastest piston-engined bombers. *(Saab)*

the first flight took place. The first production aircraft followed in August 1943, 30 months after project go-ahead. This was, of course, a splendid performance by an engineering team of less that 100 people. The production was based on a bold scheme involving as many as 500 companies, most of them outside the already existing aircraft industry. The organization and functioning of this largely inexperienced production team, however, proved to be too great a challenge. Eventually, employment in the Air Board plant responsible had to be dramatically increased to nearly 800 in order to cope with production schedules and quality requirements.

The J 22 was produced by FFVS – Flygförvaltningens Flygverkstad i Stockholm. 180 aircraft were delivered by FFVS before May 1945, sufficient to equip three fighter Wings (9 squadrons). A last batch of 18 was completed by the then new central maintenance facility at Arboga (CVA) and delivered to a reconnaissance squadron suitably equipped.

Yet another stop-gap programme – although on a much smaller scale – completes the picture of the wartime Swedish aircraft industry. Facing an acute shortage of primary trainers, the Air Board in 1942 decided to award a contract to a company at Örnsköldsvik in the north, Hägglund & Söner, to produce under licence 120 of the German Bücker Bü 181 Bestmann (Sk 25) aircraft. These were delivered between 1943 and 1946. Of wooden construction, the Bestmann was not suitable for all-metal oriented Saab which was also overloaded with work on combat aircraft. It is interesting to recall that the Bestmann was the last of the many Bücker trainers having the Swedish engineer A. J. Andersson as chief designer before he returned to Sweden in 1939. The Sk 25 became Hägglund's only aircraft venture.

The Race for More Power

For several years the Air Board had been looking for engines with more power than the Twin Wasp. After the failure to obtain the Bristol Taurus and the Pratt & Whitney Double Wasp, the German Daimler-Benz DB 601 became the centre of interest. The initial Swedish approach in the summer of 1940 met with little German interest. At the end of August the Germans informed the Swedish representatives that 'the German industry was not prepared to provide the necessary assistance for licence manufacture'.

In late 1940 new contacts were made with the Germans, with Söderberg visiting Berlin in December. There he was told by the Air Ministry (RLM) that Reichsmarschall Göring had given the following message: 'Despite the hostile attitude of the Swedish press, the Führer has decided to release a manufacturing licence for the Daimler-Benz engine (pause) on condition that Sweden delivers 100 engines per month to Germany'. The initial hopes of Söderberg faded rapidly.

At home the reaction of the Swedish Government was quite sour and the Air Board was criticized for even discussing compensations. In any case, the Air Board was instructed to give the Germans a polite no 'in view of the political and industrial implications of such a scheme, etc'. Strangely enough the Germans did not seem all that upset and some contacts were maintained. Further discussion planned for March 1941, was, however, cancelled by Berlin for unknown reasons.

In June, the Germany Embassy in Stockholm informed a somewhat surprised Söderberg that the newer and more powerful DB 605B of 1,475 hp had actually been released for Sweden, not the DB 601 which had been the subject of earlier talks. The compensation requirements were deleted completely.

In August 1941, Söderberg and his team of technical and legal experts was authorized to sign an agreement with Daimler-Benz. The negotiations with the company went smoothly until the matter of licences from major sub-contractors came up. These were 44 in number. Finally RLM intervened and issued orders to Daimler-Benz to negotiate with Sweden, One sub-contractor, Bosch, was quite impossible and told the Swedes 'that they took no orders from RLM'. They even stipulated that the Swedish companies, Hesselmann for the fuel-injection pump and Asea for the electrical equipment (ie the companies best suited to cope with the job), could not be used. But eventually, even this problem was solved.

The licence-manufacture started with excellent assistance from the Germans but in fact the engine was not ready for licensing. Several thousand changes had to be made which created serious problems for the engine industry, the Air Force units and the maintenance works. Therefore the whole programme was seriously delayed and no aircraft powered by the DB 605B could be delivered before the end of the war.

A Very Unconventional Fighter

Now that, finally, an engine in the 1,500 hp class had became available, Saab resumed project work on a fighter aircraft which had been part of the 1940 Basic Agreement but not proceeded with owing to the uncertain engine availability. During 1941 different projects were discussed between Saab and the Air Board. Air Force demands were high on performance, armament, pilot visibility and manoeuvrability. In April 1941 Saab submitted a very unconventional design, the L-13, later known as the J 21, with the engine behind the pilot and driving a pusher propeller between two tail booms. The armament was heavy and largely concentrated in the nose. But there were many technical question-marks especially as no similar aircraft had ever been developed by the major powers. One problem was the safety of the pilot in the event of a bale-out. Other problems included engine cooling on the ground as there was no propeller slipstream. Nevertheless, the J 21 design problems were solved, and the aircraft eventually became one of the first aircraft in the world to be equipped with an ejector seat. It first flew in July 1943 and went into service in late 1945. A total of 302 aircraft were built, including 183 fighters and 119 aircraft in a special attack version (A21). Yet another project based on the DB 605B was the twin-engined fighter and dive-bomber, known as the Saab 24. Initiated in late 1941, the design work proceeded all the way through extensive wind-tunnel testing

A very unconventional fighter, the pusher-propeller Saab 21A, made its first flight in July 1943. It was one of the first aircraft in the world to be equipped with an ejector seat. All J 21As were built at Saab's Trollhättan factory. *(Saab)*

to a full scale mock-up. The Air Force contract was however terminated in December 1943 following a change over from twin-engined fighters.

Both the Air Force and the industry were, of course, frustrated by being to heavily dependent on foreign powers for engines. The Basic Agreement from 1940 also, in fact, included Swedish-developed engines. At the end of 1941 therefore Svenska Flygmotor (SFA) was awarded a contract to develop an engine known as the Mx. Specified power was as high as 2,200 hp. Before the development was completed, however, the jet engine emerged both in Germany and Great Britain and in 1945 work was terminated, and the Mx design group was given new and more exciting work.

A Major Accomplishment

During the Second World War the Swedish Air Force grew from 7 to 17 Wings and the number of aircraft was almost quadrupled. The latter fact was due not only to increases in the number of Squadrons but also to significant increases in the number of aircraft per Wing. Already in 1940 the number of aircraft per Wing increased from 45 to 60 as a result of greater attrition reserves.

Between 1936 and 1945 a total of 1,395 aircraft were delivered to the Air Force. Of these 473 were imported and 922 manufactured in Sweden. 765 of the latter were of Swedish design.

The build-up of a Swedish aircraft industry began in 1937 but it took nearly five years before the industry was capable of delivering in quantity modern aircraft designed in Sweden. The creation of a competitive aircraft industry during the near-isolation imposed by the war and the lack of materials and experienced people was indeed a major accomplishment by this nation of only seven million people. The most dramatic growth phase was completed in 1944 but already in 1943 Saab surpassed Bofors as the country's largest defence manufacturer as measured in number of working hours for the country's defence forces.

The actual industry manpower growth was impressive. In 1939, Saab had 1,100 employees, including 184 engineers/draughtsmen. By 1944 employment had reached 3,967 including 756 engineers/draughtsmen. By comparison, Svenska Flygmotor (SFA) in 1939 had 344 employees including 36 engineers/draughtsmen. By 1944 the total was 1,447 including 137 engineers/draughtsmen. The engine industry was, however, larger than indicated by the SFA figures alone. Major sub-contractors, such as Bolinder-Munktell and Volvo Penta, took part in both the Twin Wasp and the DB 605B programme, their billings eventually reaching 50 percent of SFA's own.

Entering the Jet Age

In early 1945 Saab started project work with the main purpose of exploring the possibilities for a quick and inexpensive conversion of the J 21 to jet power in order to acquire as soon as possible practical jet power and high-speed experience.

When the project work began the engine type was still an open issue. The Air Force Board had ordered develop-

An unconventional provision of a jet fighter was the conversion of the J 21A as the J 21R. It was the first such conversion put into quantity production. *(Flygvapnet)*

ment of two parallel 1,500 kp class (3,300 lb) class jet engine projects, one with STAL of Finspång - an experienced turbine manufacturer - the other with SFA. The STAL Skuten* was of axial-compressor design, the SFA R 102 having a centrifugal compressor. Before either company could complete development work - prototypes were running on the bench - the British de Havilland Goblin 2 was released to Sweden for licence manufac-

* Skuten is the name of a lake near Finspång, headquarters of STAL.

ture. The latter was undertaken by SFA, while both Swedish companies actually continued design work on larger engines in the 3,000 kp (6,600 lb) class intended for future aircraft. Thus the Goblin 2 was selected to power the first Swedish-designed jet aircraft, the J 21R, which made its first flight in March 1947, only a year after start of design work. With jet power the J 21's top speed increased from 640 km/h (398 mph) to 800 km/h (497 mph). The conversion eventually proved more complex than expected and instead of 80 percent commonality in airframe design, only 50 per cent of the J 21 remained when the J 21R went into production.

As a fighter the J 21R did not prove very successful. Its critical Mach number could easily be exceeded, causing the pilot to loose control. Climb performance was not acceptable for a fighter and the range very limited. This led to a decision to cut production from 120 to 60.

After one year in Air Force fighter service, it was decided to convert the J 21R into an attack aircraft (A 21R). In this role, the aircraft actually became

Dr Erik Wilkenson (LEFT), main inventor of the Saab 'toss' bombsight. First ordered into quantity production in 1942, the sighting system was produced in several versions and later exported. It was also produced under licence in the United States. Ragnar Wahrgren (RIGHT), managing director of ASJ and 1939-1950 of Saab in Linköping. *(Saab)*

very popular and carrying a special pod for eight machine-guns – giving a total of 13 guns – became a really powerful weapons system: one of the best weapons platforms ever, according to some pilots.

Although generally speaking the J 21R was not an important chapter in the Air Force's history, it was of tremendous importance to Saab's development team as it provided vital and timely experience in transonic problems. Without these lessons, it is doubtful whether the far more advanced J 29 could have been developed so rapidly.

For the immediate postwar transition to jet fighters, starting in 1946, the Air

A Boeing B-17 Flying Fortress arrives at Saab after a forced landing in Sweden, and (BELOW) the same aircraft after conversion by Saab into a stop-gap intercontinental airliner for SILA (Swedish Intercontinental Airlines, a predecessor of SAS). Seven such conversions were made. *(Saab)* SILA'S Boeing B-17 (RIGHT) after conversion with the word Sweden boldly displayed on the fuselage. *(Saab)*

The Saab 91 Safir, a private-venture trainer/tourer, was adopted by the Swedish Air Force as a primary trainer (Sk 50) in 1951. Owing to lack of factory space the manufacture was sub-contracted to the De Schelde factory in The Netherlands where 120 aircraft were built before production was resumed in Sweden.

Force decided to acquire the British de Havilland Vampire (J 28) of which as many as 400 were delivered. Most of these were actually powered by Swedish-built Goblin engines.

Planning for Peace

In 1944 it became obvious that the war was coming to an end. Saab feared that military procurement was going to be drastically reduced after 1946 and early redundancies in the development departments were expected. Civil aircraft production was the natural planning aim, and following a decision to convert seven Boeing B-17 Flying Fortress bombers (force-landed in Sweden), into 14-passenger stop-gap long-range airliners for SILA, Swedish Intercontinental Airlines, the company board in February 1944 decided to go-ahead with a twin-engined short-haul airliner seating 24–32 passengers, later known as the Saab 90 Scandia. The prototype flew in November 1946. Flight testing was sucessful and in April 1948, ABA Swedish Air Lines (Swedish predecessor of SAS) order 10 aircraft, the first of which was delivered in October 1950. In addition to the Scandia, a three-seat light aircraft for touring and trainer use, the Saab 91 Safir, was developed. The prototype first flew in November 1945. The Safir became quite a success not least as a military primary trainer and was used by six air forces. More than 300 were built. But these two aircraft were not considered sufficient to fill the gap expected from reduced military orders. The Trollhättan factory employment in particular was in danger.

The Birth of a Major Industry

A market study for various civil products had also indicated the need for a Swedish built motorcar! It should have front-wheel drive to suit Sweden's climate and be powered by a simple two-stroke engine. By using the latest aerodynamic know-how it could be given high performance. The Trollhättan factory was considered adequate for an annual production of up to 3,000 motorcars. The economic break-even point was given as 8,000 units. Funds were made available for manufacture of three prototypes as well as for purchase of sheet metal presses which were intended for both car bodies and jet engine sheet metal parts! A production prototype was unveiled in May 1947 following more than a year of preliminary prototype testing.

The Saab 92 and the subsequent long line of successful Saab motorcars rapidly grew into a major industry in its own right, but it was born in the aircraft industry which also built the prototypes. In 1987 the Saab Car Division produced approximately 130,000 high-performance luxury cars, 75 percent of these for export. The main factory is still at Trollhättan.

'Rolling' Seven-year Plans

The Swedish politicians at an early stage realized that long-term planning was the only way to develop and produce modern aircraft, and both the 1936 and 1942 parliamentary defence decisions approved five-year planning periods for Air Force procurement. As

The Saab 90 Scandia twin-engined airliner could accommodate 24 to 32 passengers. SE-BSB is seen in the livery of Scandinavian Airlines System (SAS) in 1950. *(SAS)*

The swept-wing Saab 29 represented a technological breakthrough for the Swedish aircraft industry. This picture shows well the high surface finish of the wing. The J 29 went into service in early 1951 and for several years was the most modern fighter of European design. *(Saab)*

it turned out, even five years was a short time for complex development programmes in particular. To compensate for the disadvantages of a planning period bound to certain calendar years, in 1945 the Air Force devised a 'rolling' seven-year programme pushing the planning forward one year at the end of each fiscal year. For funds not formally committed by Parliament, a spending authorization was obtained. In the defence decision taken by Parliament in 1948, the same type of planning was also introduced for the other two Services.

Even if the shooting war had stopped in Europe, the 1948 defence decision represented a major expansion of the Air Force fighter strength, and instead of the near idle aricraft factories feared in 1944–45, the 1949 Basic Agreement between the Air Board and the industry was going to demand the highest production commitment outlined in the Agreement. No fewer than 600 of the new J 29 fighters were going to be produced over a few years!

In this situation it quickly became obvious that Saab's production capacity was insufficient to handle both the

Final assembly in Linköping in November 1949 of Saab 90 Scandias and Saab 91 Safir trainer/tourers. *(Saab)*

J 29 production and the Scandia airliner which seriously competed for both manpower and factory space. In early 1950 the Air Force suggested that the Scandia programme be terminated and finally convinced Saab by offering financial compensation, largely in the form of a bonus for each new jet fighter delivered on schedule. The 'package' also included the cancellation of 60 J 21Rs. The remaining Scandias on order were sub-contracted to Fokker in The Netherlands.

New Saab Management

In 1950 Tryggve Holm succeeded Ragnar Wahrgren as managing director of Saab. In 1950 Saab was not operating very profitably and one of Tryggve Holm's less popular initial responsibilities was to cut the shareholders' dividends. The main reasons were the heavy investment and – so far – limited returns on all three civil product programmes. Tryggve Holm's image quickly improved, however, when the new fighter, the J 29, went into large-scale production.

The J 29 had made its first flight in September 1948, only 18 months after the J 21R. In contrast to its predecessor, the J 29 embodied the latest design and manufacturing technology available. It featured swept back low-drag wings and, equipped with the powerful 2,270 kp (5,000 lb) dry thrust de Havilland Ghost engine (for which a licence was obtained, the aircraft reached a top speed of 1,035 km/h (643 mph) making it one of the fastest fighters in the world. Production deliveries started in May 1951. Also from a production engineering point of view the J 29 introduced many novelties. The high performance of the aircraft required a completely new standard of external smoothness. For aerodynamic reasons, no rivet head or sheet metal edge was allowed to project more than 0.01 in (0.2 mm) above the surrounding contour. Rivet heads were milled off. The manufacture of high-precision, high-performance aircraft was greatly facilitated by the development of Saab's system for mathematical determination of lines. This system (evolved by N. Lidbro) eliminated the traditional

Tryggve Holm, managing director of Saab in 1950-67. Under his management, Saab became an important industrial enterprise. *(Saab)*

The product range at the Saab Trollhättan factory in 1948: the Saab J 21A fighter with a prototype of the company's brand new civil product, the Saab 92 car. Today, the Trollhättan factory is a highly automated production centre for cars only. About 130,000 cars were sold in 1987. *(Saab)*

High-volume production of the Saab J 29 fighter. In 1954 alone, 221 aircraft were delivered. *(Saab)*

In the early 1950s, the Swedish Air Board and Svenska Flygmotor jointly designed and developed an afterburner for the de Havilland Ghost. Designated the RM 2B it increased the static thrust by 30 percent to 2,800 kp (6,167 lb). A total of 390 engines was modified to RM 2B standard. *(Svenska Flygmotor)*

and less reliable full-scale lofting technique and was first employed on the J 29, the shape of which was mathematically determined to approximately 85 percent (later aircraft used it to 100 percent). The system made it possible to determine exactly a very large number of reference points (co-ordinates) on the aircraft – in some cases up to nearly 500,000 – which were used throughout the design and production process. Saab also developed a method of transferring the co-ordinates to jigs and templates with great accuracy using especially developed co-ordinatographs of a type similar to that used in map-making. The high-precision manufacture also enabled Saab to comply with the Air Force's difficult requirements for 100 percent interchangeability of major parts between individual aircraft.

The production resources of the company were greatly expanded for the J 29 programme. The production rate was impressive by any standard. Between 1951 and 1956 no fewer than 661 aircraft were delivered. During September 1954 one aircraft a day was completed.

The J 29 proved to be a versatile and rugged aircraft and six different versions were developed. The last one, the J 29F, featured a Swedish-designed afterburner (there were no British) for its British-designed engine, increasing the thrust and significantly improving climb performance. The J 29 represented a real breakthrough for the Saab/Air Force team and international attention was not diminished when in 1954 and 1955 Sweden broke two world speed records previously held by the United States.

In 1950 Saab's total annual sales were only 40 million Swedish Crowns; in 1960 the figure exceeded 500 million and four years later 1,000 million. Under Tryggve Holm's forceful management, Saab rapidly became an important industrial enterprise of international repute.

1,000 Combat Aircraft

With the J 29 in full service in the mid-1950s, the Swedish Air Force reached its highest numerical strength ever; 50 Squadrons with approximately 1,000 combat aircraft.

Typical of the farsighted planning of the Air Force Board and the industry was the fact that already in the late 1940s, when piston-engined aircraft still dominated the international bomber/attack/reconnaissance scene, Sweden selected a very advanced solution for its attack aircraft programme for the 1950s.

Following extensive project studies of both twin- and single-engine configurations, on 20 December, 1948, the Air Force Board ordered development of a new two-seat single-engine transonic combat aircraft for attack, reconnaissance and night fighter use. Known as the Saab 32 Lansen (the Lance), it was the first Swedish aircraft with built-in surveillance and attack radar, indeed the first Swedish 'systems' aircraft. It made its first flight in November 1952.

Lansen became a very popular, reliable and versatile aircraft and still survives in the Air Force organization in 1988, not as a combat aircraft but for electronic countermeasures training.

A Setback for Swedish Jet Engines

Lansen was orginally planned to be powered by the Swedish STAL Dovern II (RM 4) axial-flow turbojet of 3,300 kp (7,300 lb) dry thrust which had been selected for development rather than the Flygmotor R 201 centrifugal-compressor engine. It was indeed a delicate situation as the Air Board had to inform Flygmotor in mid-1949 that it was to collaborate with its competitor STAL in the manufacture of the Dovern and the much more powerful and advanced Glan* engine project with an estimated afterburning thrust of 7,000 kp (15,500 lb) intended for a planned supersonic fighter. For STAL these projects also had considerable civil importance as the Glan was to serve as a testbed for projected stationary gas-turbine powerplants. The collaboration between the two companies did not prove easy and to facilitate the work Major General Nils Söderberg left the Air Board in order to co-ordinate the integration of the STAL development work with Flygmotor's production assignment. In July 1952, the ground testing of the Dovern had been completed at specified thrust, and flight testing started in a modified Avro Lancaster bomber. Then in November 1952 there was a sudden bombshell: the Air Board had decided to abandon both the Dovern and Glan following a recent offer from the United Kingdom for a previously refused licence for the Rolls-Royce Avon of similar thrust. The financial terms were extremely favourable and Sweden was offered full access to future Avon developments.

*Dovern and Glan are lakes near Finspång.

The Saab 32 Lansen all-weather attack aircraft was Sweden's first 'systems aircraft'. It featured a built-in radar mainly for use with the revolutionary RB 04 radar-homing anti-ship missiles. *(Flygvapnet/F 15)*

More than 400 Lansen aircraft were produced. *(Saab)*

Many people, of course, regretted the decision, which technologically was a blow to the promising Swedish jet engine industry. On the other had it may be argued that with the limited funds available to the Air Force, the development costs for a line of Swedish jet engines with ever increasing complexity would have made it extremely difficult to finance in the long term indigenous engine development and production. For the Air Force the decisive factor was that the Swedish industry remained capable of licence manufacturing and support of the engines. For several decades now the industry has actively participated in the development of the engines being built under licence and also contributed with quite a number of advanced afterburner designs to meet special Swedish requirements. Today, the Swedish engine industry is also an important partner for both development and production programmes for large US commercial jet engines.

Despite this disappointment for the STAL company and for Curt Nicolin, the chief engineer leading the development team, the plans for a line of stationary gas-turbines based on the Glan jet became a reality and the 10,000 kw GT 35 eventually sold to several countries. For Svenska Flygmotor, the Air Board decision led to production of more than 1,100 Avon engines (RM 5 and RM 6). The Avon powered Lansen right from the first flight.

A Missile Parenthesis

The first German guided missile, although it was then called in Sweden an aerial torpedo, landed in Sweden in error. In November 1943 a V 1 test vehicle crashed near Karlskrona in southeast Sweden. Only two weeks later another V 1 was found near Ystad. In 1944, two more missiles crashed in Sweden. The first one was yet another V 1 but the second proved a sensation, - a V 2. A very detailed examination was made and was in fact made available to the Allies by way of Britain in return for some radar equipment. All three Swedish military Services were of course excited by the new weapon technology unveiled. The Navy looked at the guided weapon as a complement to the torpedo, the Army as an alternative to anti-aircraft artillery, and the Air Force as a pilotless aircraft for various types of mission.

A small central research and test organization was formed in February 1945. The private companies, mainly Saab and STAL, were given the task of examining the German V 1 test vehicles under contract to the Navy. The two companies designed and manufactured on a limited scale a number of experimental missiles based on V 1 technology with STAL-built pulse-jet engines but later also with a Swedish-developed expendable jet engine. The first Saab-designed missile, the Rb 310, made its first flight in 1947 and was produced in a small series. It was successfully used as a target missile for various purposes. A larger version, the Rb 311, was also flight tested. A third, much larger surface-to-surface missile, the Rb 312, was designed by Saab under contract to the Air Board and was to be powered by a Swedish jet engine developing a thrust of 740 kp. The Rb 312 project was, however, terminated before flight testing could start because of autopilot problems.

In April 1948, a joint inter-Service missile bureau (Försvarets Robotvapenbyrå) was formed under the Air Force Board. At this time most of the missile development and manufacturing activities were transferred from private industry to government facilities. The central Air Force maintenance works, notably the new Arboga facilities (CVA) with underground workshops, played a leading role. In 1949, CVA started its missile activities by building a second series of the Rb 310 target missile.

In 1950 development started on the Rb 304 radar homing anti-ship missile which was one of the main weapons on the Saab 32 Lansen all-weather attack aircraft. With the Rb 304 Sweden became a pioneer in anti-ship missiles, and it was produced in quantity by CVA during the period 1958 to 1964. From 1949 Saab was almost out of missile activities for nearly 10 years, but there was to be a return to this activity.

A New Industry Team Emerges

The Second World War isolation was a major challenge for the organization of a network of equipment and materials suppliers to the aircraft industry. It took years to familiarize the suppliers with the special aircraft quality requirements but in the end the companies involved realized that by working closely with the aircraft industry they learned how to tackle many technical challenges, however difficult a customer Saab seemed at times to be. Despite the hundreds of suppliers engaged by Saab, the company had to undertake design and manufacture of a lot of mechanical and electrical equipment which simply could not be bought

Banker Marcus Wallenberg *(right)*, **Saab Board chairman 1968-1983 and a member of the Board ever since 1939, seen here inspecting Draken final assembly line with Tryggve Holm.** *(Saab)*

The assembly facilities at Linköping were further expanded in 1953-55 to accommodate Lansen production. *(Saab)*

from outside. In the 1940s the largest single equipment item designed and produced within Saab was the bombsight first produced in 1942. Saab also designed and manufactured the undercarriages, control systems, navigation equipment, etc for both aircraft and missiles. Most of the in-house equipment manufacture took place in a section of the underground factory completed in Linköping for the purpose, and in 1957 a major new expansion was inaugurated. The reason was that aircraft had become more and more complex. This trend was already apparent in the late 1940s and particularly affected the development of the Saab 32 Lansen, which marked Saab's entry into the electronic age. The great technical challenge was the development of the airborne radar which involved a number of companies such as LM Ericsson, SRA, Saab, and CSF in France, Saab, AGA, and Philips were engaged in the development of sighting systems and autopilots, gyros, communication radio and navigation systems. Countermeasures remained an Air Board responsibility.

The first of a line of fighter aircraft sighting systems developed by Saab was the S 6A. Like the SA 04 autopilot, also for Lansen, it was produced by the Jönköping factory. A new Swedish aircraft industry teamwork had started.

For the new generation of supersonic fighter aircraft in the planning stage in 1950, even more advanced electronic systems were needed, both in the air and on the ground. To meet the new technical challenge, in 1954 Saab decided to create a special department for operations analysis and systems engineering employing some 50 qualified engineers. This important upgrading was largely due to the initiative of Dr Tore Gullstrand, later general manager of Saab-Aerospace.

A Qualified Customer

In military aircraft development it is, of course, tempting to choose conventional solutions in order to reduce technical risks and costs. But such a policy seldom produces winners, and the fighter pilot knows only too well what a performance or manoeuvrability advantage means in the 'acid test', the air combat duel. At least this was, somewhat simplified, the situation in the 1950s and 1960s and to a high degree even today although 'systems performance' has now come to dominate the discussion. However, for a multi-role aircraft armed not only with guided missiles but also with cannon and other conventional weapons, highest possible 'platform' performance is still equally decisive.

A small nation such as Sweden cannot afford conventional or, for that matter, too specialized aircraft. To compensate for limited quantity, the quality must be as high as technically and economically possible.

The success of an aircraft industry is highly dependent on the ability of the customer, the Air Force, to define early enough the threat and specify the requirements to meet it. Sweden has been lucky in this respect. Through close teamwork between the government organizations and the industrial companies involved, unique solutions to truly difficult problems have been found and, what is equally important,

The Saab management team at Linköping in front of the Saab J 35 Draken, Sweden's first supersonic fighter, in 1955. (*Left to right*): **Bengt R. Olow, chief test pilot; Erik Bratt, project manager; Tryggve Holm, managing director; Lars Brising, technical director; Tord Lidmalm, chief engineer design departments; Hans Erik Löfkvist, chief engineer technical analysis department; and Kurt Lalander, chief engineer test departments.** (*Saab*)

Draken - Sweden's first Mach 2 aircraft, with its trendsetting double-delta wing. *(Saab/I.Thuresson)*

Every Swedish Air Force fighter pilot was trained to operate Draken from road bases. *(Flygvapnet/F4)*

the government planning has been firm. Only by a consistent policy has Sweden been able to produce the combat aircraft needed for its security policy.

The Swedish Air Force leaders as well as the defence politicians deserve praise for understanding the industrial problems involved.

The foregoing may serve as an introduction to what is perhaps the most daring chapter so far in the history of the Swedish aircraft industry, the Saab Draken supersonic fighter. Here again planning began very early, in fact only a year after the first flight of the J 29. In October 1949 a first contract was let to Saab for a design study for a project known as the Aircraft 1200. Two parallel projects were pursued, one for a conventional swept-wing aircraft and one using a delta-wing configuration. In the latter project the Saab design team used a completely novel double delta aerodynamic configuration combining the advantages of the delta wing with those of the swept wing and elimina-

The sale of the first of fifty-one Draken attack, reconnaissance and trainer versions to Denmark in 1968 represented a major export breakthrough. *(Saab/I. Thuresson)*

ting the disadvantages of the two, giving optimum performance and flight characteristics. After much research work and by the building and flight testing of a 70 percent scale research aircraft, the double delta was finally chosen in early 1952. The first prototype flew in late 1955 and the first true Swedish supersonic fighter was a reality.

The J 35A Draken went into service in late 1959 and was followed by five different versions for the Swedish Air Force including the J 35F which provided Sweden with the most efficient all-weather fighter of European design for many years.

An Export Breakthrough

Draken soon became the subject of considerable foreign interest. In 1968 the Danish Air Force ordered the first

The S 35E reconnaissance version of Draken was unarmed and relied on its supersonic speed, even at low altitude, to avoid interception. *(Saab/I. Thuresson)*

twenty of a total of 51 Draken aircraft in two new long-range versions for attack and photographic-reconnaissance plus two-seaters for training. In 1970, Finland ordered the 35S, a version of the J 35F which was assembled in Finland, and in addition, has since ordered a number of ex-Swedish Air Force Drakens. In 1985, Austria ordered twenty-four refurbished ex-Flygvapnet J 35Ds (35OEs in Austria). A total of 604 Draken aircraft were built between 1955 and 1972.

Back into the Missile Business

In 1961 a new Basic Agreement between Saab and the Air Board was signed. For the first time it included the production of guided missiles, the Rb 27

In 1970 the first Draken fighters were sold to Finland. *(Saab/I. Thuresson)*

More than 600 Drakens were built, the last one in 1972. *(Saab/I. Thuresson)*

and Rb 28 (Hughes HM-55 and -58) air-to-air weapons which were ordered in large quantities. Deliveries started in 1962. The licence production was a major project involving considerable investment both at Linköping, where final assembly of both versions took place, and at Jönköping where 'super clean' rooms were built for the homing heads. A special warhead final assembly and check-out facility was also built at Linköping.

In the mid-1960s a new air-to-surface missile was developed by Saab, the Rb 05, featuring a radio-command steer-

The Swedish Air Force uses many simple wooden hangars for protection of ground crews and equipment at turnaround sites. Here a J 35F Draken is being readied for another mission. *(Saab/I. Thuresson)*

Starting in 1962 Saab produced a large number of American Hughes Falcon air-to-air missiles under licence for the J 35F version of Draken. Both the IR-homing HM-58 (Rb 28 in Sweden) and radar-homing HM-55 (Rb 27) were manufactured. *(Saab/I. Thuresson)*

ing system. Powered by a Flygmotor-designed liquid-rocket engine, the production version became operational in 1972.

A very different type of missile, the Rb 08, went into production at Saab in 1967. It was a Swedish development of the French jet-powered Nord CT-20 target drone designed to the requirements of the Swedish Navy for both ship-to-ship and coastal purposes.

In 1969, Saab was contracted to develop a modernized version of the Rb 04 which had been in use with the Air Force since 1960. The new version was designated the Rb 04E and includes a new rocket engine.

The considerable experience that Sweden has acquired over the years, notably in anti-ship missiles, was further developed in the late 1970s. Based largely on the Rb 04 experience, in 1977 Saab submitted a project to the Navy for a new anti-ship missile to be based aboard the Navy's fast patrol boats. Designated the RBS 15, it was ordered into full-scale development and production in 1979. The missile is powered by a jet engine (plus booster rockets), has a range of approximately 70 km, and like its predecessors it is a sea-skimmer. The RBS 15 was operational from 1984. Later, the Navy acquired the missile for its new coastal corvettes and also ordered development of a vehicle-based coast defence version. The missile system has also been exported to Finland.

Saab's missile activities are now concentrated within a separate subsidiary company, Saab Missiles AB.

A considerable number of missiles were launched from Draken to verify the performance of the Rb 27/28 family. The number of launches is indicated on the fin. *(Saab/I. Thuresson)*

Apart from naval applications of the RBS 15, the company is now developing an air-launched version for the Air Force, the RBS 15F. Externally it is identical to the Naval versions but no booster rockets are carried.

Sweden's missile procurement has, however, not been characterized by the same firm 'family' planning as aircraft development and procurement. Although several advanced Swedish missile have been developed, including air-, ship- and shore-launched anti-ship weapons as well as short-range laser-homing anti-aircraft missiles (Bofors RBS 70) and anti-tank missiles (Bofors RBS 56 Bill), no Swedish-designed air-to-air, ground-to-air or long-range surface-to-air missiles have reached quantity production. A promising IR-homing air-to-air missile, the Saab-designed Rb 72, was cancelled in 1978 for financial reasons. It was to be replaced by the United States AIM-9L Sidewinder (Rb 74) although deliveries of the latter could only start in 1987.

At the present time, the Swedish

The Rb 08 was a ship- and shore-based anti-ship missile system produced in the 1960s by Saab for the Swedish Navy. *(Saab)*

Government has allocated considerable funds for a joint Swedish-British programme for a new active-radar homing development of the British BAe Sky Flash air-to-air missile, an earlier version of which is already in service with the Swedish Air Force as the Rb 71. Other types of Swedish missiles are in the planning stage.

Mainly to co-ordinate marketing activities, in 1978 Saab and Bofors joined forces in a separate subsidiary company, Saab-Bofors Missile Corporation (SBMC).

A Civil Project Turned Military

In 1958 the company management again feared that it would take some years before a new generation of combat aircraft could be developed for the Air Force although some preliminary project work had already started. In order to bridge the expected gap in the workload for the development departments, some project and market studies were made for a small high-speed executive jet transport. One project was for a twin-jet delta-wing aircraft. Market research did not produce too promising results, however, and instead a new military project for a trainer/liaison aircraft, the Saab 105, was launched as a private-venture in April 1960. Then the main intention

During the late 1960s Saab developed the Rb 05A radio-command guided air-to-surface missile which has been in Flygvapnet inventory since 1972. Most verification trials took place from a two-seat Draken trainer. *(Saab/I. Thuresson)*

Sweden was a pioneer of air-launched anti-ship missiles. The picture shows the Rb 04E improved version first produced in 1969 and still equipping Flygvapnet attack squadrons. *(Saab/Å. Andersson)*

The Saab RBS 15 is the standard ship-to-ship missile in the Swedish and Finnish Navies. This photograph shows a launch from a Swedish Spica II fast missile craft.

was to meet a Swedish Air Force requirement for a new combined trainer/light attack aircraft.

Later in 1961, the Government authorized financing of the development work, and in early 1962 the Air Board signed a preliminary agreement for procurement of 130 aircraft on condition that the 'aircraft fulfilled the requirements'. Eventually it did. Powered by two French Turboméca engines, it flew exceeds 20 years. This has been made possible through on-going modernization (regular inspection and sometimes modications) but above all to higher component reliability and maintainability. In reality a new maintenance philosophy has been developed, keeping track of major component's lifecycle and MTBF (mean time between failure rate). The considerable period of time between each completely new type that existing aircraft should be equipped with new weapons and fire control systems. This was accepted and led in late 1958 to the development of the J 35F but also delayed the upcoming decision on the new generation of aircraft. The project studies actually went on until 1961 when the Supreme Commander approved the operational requirements (specification) for a new multi-role aircraft to replace Lansen

The Saab 105 undergoing final assembly at Linköping. This aircraft was the first of forty delivered to Austria starting in July 1970. *(Saab/Å. Andersson)*

for the first time in July 1963. A total of 150 aircraft of different versions were later supplied to the Swedish Air Force. In 1967, Saab developed, again as a private venture, a more powerful version of the Saab 105 equipped with two General Electric J85s with 70 percent more thrust and much higher performance. In 1968, Austria ordered the first of a total of 40 such aircraft.

Careful Considerations

In 1942 the life of a combat aircraft (in peace time!) was set at seven years. In 1958 the authorities assumed a life of 7–8 years. For Lansen the period increased to 15 years in the early 1970s and for the latest version of Draken it of combat aircraft is, of course, a result of the dramatically increased cost of each new system. Naturally, export is becoming more and more important.

Project studies tend to start very early indeed. The decision to go ahead with development of the J 35 Draken was, for example, taken in 1952. Immediately thereafter a series of project studies were initiated regarding the next generation of aircraft for the Air Force. Four totally different families of fighter, attack and multi-role aircraft were studied and balanced against on-going studies of air defence missiles (the international debate regarding manned aircraft v. missiles was, of course, felt also in Sweden). Not surprisingly, the study recommended (A 32, S 32) attack and reconnaissance aircraft as well as, later on, the Draken family.

Aerodynamically, Viggen, as the new aircraft was later named, proved to be extremely advanced and truly unconventional. For the first time in the world, a nose wing (canard) was used in combination with a delta-shaped main wing. The configuration provided outstanding lift characteristics, giving short take-off and landing distances and thus reducing vulnerability on the ground without having to resort to expensive and complex swing-wing configurations or lift/deflected thrust engines.

Viggen fighter version taking off from a road base. Despite a weight of 18 tonnes (39,683 lb), it can take off and land in about 500 m (1,640 ft). *(Saab/H. O. Arpfors)*

Sweden's Largest Defence Project Ever

In April 1962 the Air Board informed the Swedish aircraft, engine and equipment industry that for the Viggen programme a new procurement system using a main contractor would be applied. In a later instruction, the responsibilities of the Air Board, the main contractor, associate contractors, and sub-contractors were defined in detail.

In 1964 another important change took place in the organization of the armed forces equipment procurement. The Service Chiefs were no longer in command of their technical/economic Boards. Instead, the Boards were made directly responsible to the Government (Defence Department), in an obvious effort to achieve better control of the expensive equipment procurement programmes.

In 1968 this move was followed by an amalgamation of the Services Boards to one authority, the Defence Material Administration (FMV). For Air Force matters, an Air Material Department (AMD) was formed inside FMV.

In its new role Saab was not only responsible for the development and integration of the basic aircraft, it was also contracted to develop its central digital computer. The central computer (CK 37) which greatly facilitated the pilot's work – Viggen was a single-seater compared to the two-seat Lansen it was going to replace – was developed from a 'breadboard' prototype, the D 2, which the Air Board had already ordered in 1958 as a technology programme. The D 2, incidentally, not only led to the CK 37 which was ready for testing in 1963, but also to a new Saab civil product line, general-purpose computers. The first such computer, the Datasaab D 21, was delivered to a customer in 1962. Technically it compared well in reliability with the contemporary IBM systems. The general-purpose computer market eventually became extremely competitive and in 1980 Saab decided to sell off its computer and related activities to L.M Ericsson, the Swedish world-wide telecommunications group.

By that time all the CK 37s for the first Viggen generation had been delivered. The Viggen development work which culminated in the 1967–71 period was the largest defence project ever undertaken in Sweden. More than 3,300 engineers were employed by the Swedish companies involved in the project, and, all told, the development and production involved more than 10,000 people.

The Viggen development team also included several important new foreign members, the most important being a new engine manufacturer. After more than 10 years of the Rolls-Royce Avon series for Lansen and Draken, the United States Pratt & Whitney JT8D by-pass engine was chosen to power Viggen. The engine was completely built in Sweden under licence, with Flygmotor also playing a major role in the transformation of this subsonic airliner engine into an afterburning, supersonic engine.

Not surprisingly, a project of this magnitude drew criticism, not least from the extreme Left which, for some reason, seems to have easy access to newsmedia. For this reason, in 1972 the Government published a 'white book' which outlined in great detail the different study and evaluation phases, the requirements, the choice of major subsystems, the development work and the production preparations until the formation of the first combat unit in 1972.

Viggen made its first flight in 1967 and the first production aircraft (AJ 37) appeared in 1971.

After the AJ 37, the Sk 37 two-seat conversion trainer followed in 1970, the SF 37 photographic-reconnaissance and the SH 37 sea-surveillance versions in 1973, and in 1974 the first JA 37 all-weather fighter version. Although the JA 37 looks virtually the same as the earlier Viggen versions it must be regarded as almost a new aircraft system incorporating much of the technical development which took place since the AJ 37 specification was approved in 1961. The modernization was necessary because of new threats and includes a more powerful engine, new avionics and, of course, new armament.

Saab Remodelled

As a company, Saab was considerably remodelled during the 1960s. It had grown tremendously, particularly automobile production and computers and related activities (bank terminals, minicomputers, etc). Its registered name, Svenska Aeroplan Aktiebolaget (SAAB) no longer properly described its products and in 1965 the name was changed to SAAB Aktiebolag. The company's name could also be written as Saab.

In 1967, Tryggve Holm stepped down as managing director, being succeeded by Dr Curt Mileikowsky. Holm remained on the board.

In 1968 a major development was the announcement of a merger between Saab and Scania-Vabis, the major truck manufacturer with headquarters at Södertälje south of Stockholm. Scania-Vabis belongs like Saab to the Skandinaviska Enskilda Banken (Wallenberg) group and it was therefore decided to merge the two companies in order to make better use of the joint resources for research and development, production and export marketing in the automobile sphere. The new company was later registered as SAAB-SCANIA Aktiebolag.

The merger also brought about great

In April 1965 a very detailed design mock-up of the Saab 37 Viggen supersonic multi-role combat aircraft was unveiled at Linköping. In the Air Force/Saab team line-up are: Major General Greger Falk, head of the Air Board; Tryggve Holm, Saab managing director; Lieutenant General Lage Thunberg, C-in-C of the Air Force; Colonel Åke Sundén, head of the Air Material Department in the Air Board; and Lars Brising, Saab technical director. *(Saab/Å. Andersson)*

Viggen camouflaged. The shadow shows the double-delta layout. *(Saab/Å Andersson)*

The development of the Saab 37 Viggen was a bold enterprise by the industry and the Air Force. The aerodynamic configuration was extremely advanced using a canard fore-plane in combination with a delta-shaped main wing. Viggen was also the first single-engined aircraft incorporating a thrust reverser. *(Saab/I. Thuresson)*

In September 1974 the Defence Ministers of four NATO countries, Belgium, Denmark, The Netherlands and Norway, visited Sweden to view the Saab 37 Viggen. In this picture Dr Tore Gullstrand, general manager of Saab-Scania Aerospace *(left)*, is seen with General Sverre Hamre, Norway, chairman of NATO's Steering Committee (for the Starfighter replacement programme), Dr Curt Mileikowsky, Saab-Scania president, and the Defence Minister of The Netherlands Henk Vreedeling.

changes in the organization of the company, and activities were mainly concentrated into four profit centres of which the Aerospace Division was one. As its first general manager, Dr Tore Gullstrand was appointed.

Through an acquisition in 1968, the Aerospace Division also included Malmö Flygindustri (MFI), a small manufacturer of light aircraft including the MFI-15 piston-engined trainer which made its first flight in 1969 and later went into quantity production, mainly for export. It has also been produced under licence in Pakistan.

The new Saab-Scania organization created in 1969 had nearly 27,000 employees. In the early 1970s the company expansion continued in the automotive field and during 1975 employment reached 35,000 of which 5,000 worked in the Aerospace Division. The

Historical Survey 53

Viggen final assembly. The vertical fin can be folded allowing the use of low hangars and underground shelters. *(Saab)*

company's total sales reached 7,900 million Swedish Crowns of which the Aerospace Division was responsible for 11 percent or 858 million. Although limited in size, the Saab-Scania Aerospace Division enjoyed a position of unusual financial strength among the European aircraft manufacturers. A highly profitable and well diversified group, Saab-Scania generated in 1975 an operating profit of 365 million Swedish Crowns after depreciation etc. The share capital had reached 524 million Swedish Crowns and the number of individual shareholders 53,000. On the management side, in May 1978 Sten Gustafsson succeeded Curt Mileikowsky as managing director.

Viggen Export Efforts

Following the 1968-70 successful sales of Draken and the Saab 105 to

In 1969, Saab merged with Scania-Vabis and became Saab-Scania. The company headquarters are at Linköping, also centre for aerospace activities. *(Saab)*

Denmark, Finland and Austria, considerable effort began in the early 1970s to market Viggen. This began in earnest in 1970 when a questionnaire arrived from Australia and led to an outright evaluation in 1972. A change in the Australian Government delayed the procurement several years, however, and eventually resulted in direct procurement from the United States of a limited number of General Dynamics F-111 long-range strike aircraft.

In 1973-74, four NATO members, Norway, Denmark, Belgium and The Netherlands, agreed to co-ordinate the procurement of a replacement for the Lockheed F-104 Starfighter. A special joint procurement organization was set up, chaired by the Norwegian Army General, Sverre Hamre. This organization was not only responsible for a joint requirement but also for organizing the production and the industrial offsets requested by the four nations.

Saab-Scania went to considerable trouble to define a joint European co-production scheme for Viggen and also offered very extensive industrial offsets outside the aircraft and related areas. In May 1975, however, the General Dynamics YF-16 was declared the winner of the competition which also included the Northrop YF-17 and the Dassault Mirage F-1. The YF-16 was a newer aircraft than the JA 37 fighter version of Viggen but it may still be argued whether the two 'weapon systems' were really comparable in 1975. In addition, it is no secret that US political pressure on the four NATO countries to select a US aircraft was considerable and difficult to match.

Yet another serious effort was made in 1977-78 to export Viggen, this time to India which made a very detailed technical and operational evaluation of the aircraft. The plans called for direct purchase as well as licence manufacture. In 1978, however, the United States refused to release the Pratt & Whitney JT8D engine to India. Such release approval was, of course, part of the original Swedish-American licence agreement.

Despite problems of a mainly political nature encountered in the export efforts, Sweden will continue to market its military aircraft (and missiles) to countries approved by the Swedish Government. The new generation of military aircraft now under development is already generating much foreign interest.

A Very Difficult Problem

During the 1960s and 1970s Sweden's defence forces were considerably restructured as a result of the reduced buying power of the defence budgets and the ever-increasing cost of modern equipment. From 1960 when the Air Force comprised more than 800 aircraft, the Service has been severely reduced in numbers. In 1986 there were only about 350 aircraft for fighter, attack and reconnaissance missions, plus five squadrons of combined trainer/light attack aircraft. The market for Swedish military aircraft has thus changed dramatically. During the 1990s a major part of the Air Force squadrons will need new equipment, notably attack and reconnaissance elements. A new combat aircraft to follow Viggen and the Sk 60 (Saab 105) in service became a hotly debated defence planning issue as early as about 1975.

After an initial study phase at that time of several completely new multi-role aircraft, the Parliamentary Defence Committee at the end of 1975 recommended development of a new version of Viggen known as the A 20 (The A designation stood for modification of existing aircraft and a B for completely new project) whereas the need for a new light attack/trainer aircraft would require further studies. The latter were subsequently concentrated on a project designated B3LA (LA for light attack). Fairly soon, however, it became obvious that the funding to be made available by the politicians would not be sufficient to finance both the A 20 and the B3LA with the original Air Force ambitions. A Government-appointed committee was unable to choose between the alternatives available and as a result Parliament in 1978 decided to allocate 310 million Swedish Crowns for further project work on both aircraft types. Late in 1978 the economic situation prevented the additional funding of 350 million Swedish Crowns annually as required in the Air Force budget to finance the B3LA. In addition, the three-party coalition government collapsed, making decisions even more difficult.

In February 1979 the new Government suddenly said no to both the B3LA (and less expensive derivatives studied) and A 20 plans. This was a major shock to the industry, notably to Saab where much design work had been completed, including a full-scale mockup of the B3LA along with advanced project work on a new missile system. A new Government committee, appointed to look into civil, that is non-aviation, applications of military technology developed in the aircraft industry, submitted its report in early 1980 but with little tangible result.

Within the Air Board new studies continued during 1979, including foreign alternatives authorized by the Government. This new move opened the possibilities for competing Swedish industry proposals, and already by the end of 1979 the Supreme Commander was informed of a new Swedish multi-role alternative, the JAS project. The Supreme Commander now recommended that planning for a new combined light attack/trainer aircraft be suspended and instead be directed towards a new multi-role aircraft – Swedish or foreign – for replacing, starting around 1990, all versions of Viggen. The Supreme Commander also recommended up-dating of three or four squadrons of Draken fighters to serve until 1995. In March 1980, the Government endorsed the Supreme Commander's proposal for a JAS project – Swedish or foreign-designed and Swedish or foreign manufactured.

A New Factor in the Swedish Aircraft Industry

The Swedish aircraft industry now formed a special industry group, the JAS Industry Group comprising Saab-Scania, Volvo-Flygmotor, Ericsson Radio Systems, and FFV, to closely co-ordinate the project and produce a joint proposal and formal offer in response to an Air Board Request for Proposals sent out in early 1981. The joint offer

In 1976-79 Saab worked on a projected subsonic light attack and trainer aircraft under the tentative designation B3LA. Originally planned alongside yet another version of Viggen known as the A 20, both the B3LA and the A 20 were cancelled in February 1979. Instead, a new supersonic multi-role combat aircraft, the JAS, was accepted in principle in early 1980. The drawing shows the B3LA two-seater which was intended to be powered by a General Electric F404 without afterburner. *(C. G. Ahremark)*

was submitted by 1 June the same year.

In the meantime, in early 1980, the Government had tentatively authorized 200 million Swedish Crowns to finance project work on condition that the industry invested a similar amount unitl 1982, a condition accepted by the industry. At the request of the Government the development work was to include more international sub-systems suppliers than in earlier projects in order to benefit from longer production runs and lower unit costs. The industry group also proposed extensive use of new technology throughout the aircraft to reduce weight and thus costs.

Whilst the initial design work continued, the negotiations between the industry group and the Air Board were completed on 30 April, 1982. Government approval followed on 6 May and on 4 June Parliament approved the new five-year defence decision which included the JAS programme. This included a series of 140 production aircraft until the year 2000, within an estimated budget of 25,700 million Swedish Crowns at the 1981 price level for development and procurement. Of this sum, the JAS Industry Group companies were responsible between them for about 16,000 million, the Saab-Scania share being about 65 percent. The JAS Group total did not include funds for weapons, countermeasures, ground support, communications and training systems, etc. Now, full-scale development of the JAS 39 Gripen (Griffin), as the aircraft was named, had begun.

The Gripen programme is scheduled to continue well into the next decade. Although so far (1988) only 140 aircraft have been authorized for production, the total Air Force requirement will exceed 300 aircraft assuming a one-for-one replacement of all Viggen and Draken aircraft in service.

A Radical New Approach

The Government's decision in 1979 to cancel the B3LA attack/trainer development programme led to a complete reassessment of the Swedish aircraft or, more correctly, aerospace industry and its future. Military aircraft represented 83 percent of the Aerospace Division's total sales in 1979, and its future workload was by no means certain. Commerical aircraft proved to be the only natural consideration. An initial step into commerical aviation had been taken already in 1979 when the company received a contract from McDon-

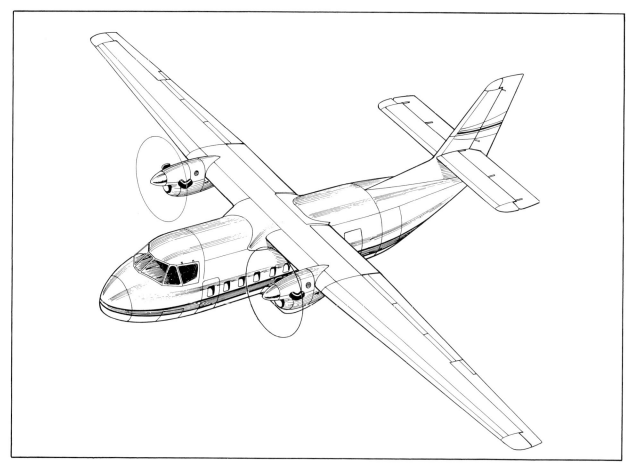

In the late 1970s, Saab worked on a number of light transport projects. In 1979, the company was technically prepared to go ahead with a twin propeller-turbine high-wing regional airliner and multi-role transport, the Saab 1084. *(C. G. Ahremark)*

nell Douglas to produce inboard flaps and vanes for the MD 80 series of jetliners. This work was later expanded, together with major sub-contracting of structural components for the British BAe 146 four-jet airliner. But now, a radical new approach was vital.

In early 1979 the Saab-Scania Board took a major decision to invest heavily in commercial aircraft development and production in order to achieve a balance between military and commercial aircraft over a 10-year period. Several years of project and market studies indicated that international collaboration would be of great advantage from many points of view. Development risks would be reduced and a greater home market provided. Later in the year, a partnership with Fairchild Industries of the United States was discussed, Fairchild already being established in the commuter/regional air-

In late 1985, Saab-Scania took complete control of the Saab 340, as the aircraft is now called. The picture shows the new commercial aircraft facilities built in 1981-82. A new building for wing manufacture was added in 1987. *(Saab)*

In 1980, Saab-Scania embarked upon a very ambitious programme for the development and production of a twin propeller-turbine regional airliner jointly with Fairchild Aircraft Industries Inc, of the United States. The 35-passenger Saab-Fairchild SF 340 went into scheduled service in June 1984 with the launch customer, Crossair of Switzerland. By the end of 1988 this airline had ordered a total of 24 aircraft. (Saab/Å. Andersson)

liner field. At the end of January 1980, an agreement was signed between the two companies for joint development, production and marketing of a commuter/regional airliner seating 34 passengers in a comfortable, pressurized cabin. The aircraft would be powered by two propeller-turbines and have a cruising speed of about 500 km/h (310 mph) and feature low noise level and, not least, good operating economy. The total world-wide market for this category of aircraft was estimated at 2,000 aircraft over the next 20 years with roughly half the world market in the United States.

The development work, which was shared between the two companies, started immediately and with all possible speed – the competition was not idle! To facilitate a rational production of this high-technology commercial aircraft which included an unprecedented degree of metal-to-metal bonding, Saab-Scania in December 1980 decided to invest more than 200 million Swe-

Georg Karnsund (left), Saab-Scania president, with Moritz Suter, president of Switzerland's regional airline Crossair, the launch customer for the Saab-Fairchild 340 in 1980. (Saab/M. Thörnblad)

Harald Schröder *(right)*, **general manager of Saab Aircraft 1983-87, discussing some aspects of the JAS 39 Gripen with Tage G. Petersson, Sweden's Minister for industry. On the left is Tommy Ivarsson, Gripen project manager and head of Saab Military Aircraft.**

dish Crowns in a completely new 25,000 sq m (270,000 sq ft) factory, and a major part of the factory was ready for use in 1981.

In October 1982, the first Saab-Fairchild SF 340 was rolled out before more than 600 invited guests, including Carl XVI Gustaf, King of Sweden, and on 25 January, 1983 – three years to the day of the agreement with Fairchild – the first flight took place.

On 30 May, 1984, the Swedish Board of Civil Aviation (BCA) issued its airworthiness certificate and one month later approval followed by the US Federal Aviation Administration (FAA) and the authorities of the 10 countries in the Joint European Airworthiness Group (JAR).

On 6 June the first aircraft was delivered to the launch customer, Crossair of Switzerland. One week later it went into scheduled service. Later in the year, the aircraft also went into scheduled service in the United States.

In the autumn of 1984 some disturbing and unrelated failures occurred with the General Electric CT7 propellerturbine powering the SF 340 which, unfortunately, led to them being grounded by the authorities. The engines were, however, rapidly modified and returned to service. By the end of 1985, 36 aircraft were in service with seven airlines on three continents.

From 1 November, 1985, Saab-Scania took over complete control of the SF 340 programme with Fairchild remaining only as a sub-contractor (for the wings and the empennage) up to and including aircraft No.109. During 1987, all wing and empennage production was transferred to Sweden where production facilities were further extended.

On 4 September, 1987, the 100th aircraft was handed over to its purchaser, Salair of Sweden. The aircraft has now been redesignated the Saab 340. The 1979 ambition of the Saab-

Scania Board, to achieve a balance between military and commercial aircraft, was actually reached in 1986 when commercial aircraft represented 52 percent of total aircraft sales which amounted to 3,276 million Swedish Crowns, more than double the figure of five years before.

By its contribution to the country's air defence over the past 50 years, the aircraft industry now continues to play a vital role in the Swedish security policy. In less than a decade, Saab-Scania has also established itself as the leading European manufacturer of aircraft for regional airline operations.

As a company, Saab-Scania has grown into a powerful industrial group. During 1987, its total sales (trucks, buses, cars, aircraft, electronics, etc) reached 41,400 million Swedish Crowns. Total employment exceeded 50,000. Aircraft sales represented 4,400 million Swedish Crowns, and employment 6,500.

An outstanding view of two B 17s of F 4 *(Flygvapnet)*

Saab 17

For the fledgling Swedish aircraft industry the Saab 17 represented a great technological challenge. It was the first all-metal, stressed-skin aircraft ever designed in Sweden.

A light bomber and reconnaissance aircraft for Army and Naval use, the design began under the project designation L-10. For Naval use, a float-equipped version was designed.

The L-10 was selected for development following an Air Force evaluation of the two different projects submitted, one by AB Förenade Flygverkstäder (AFF), the joint development and contract management company formed in 1937, the other by ASJA. A contract for two L-10 prototypes was awarded to ASJA on 29 November, 1938.

To make possible a rapid start for the development work, ASJA had augmented its engineering staff by hiring a total of 46 American designers and stress specialists in 1938–39. Their stay in Sweden, however, was to prove rather brief as most of them were called

The Saab 17 was a clean-looking design. Four different versions were produced. The B 17B powered by a Swedish-built Bristol Mercury engine was the first to go into service. It is seen here equipped with retractable skis. *(Saab)*

back to the United States when war began in Europe in September 1939. Their input of experience was very valuable, however, and significantly contributed to the excellent reliability of the aircraft. The first prototype of the Saab 17, as the aircraft had now been named, made its first flight on 18 May, 1940, with the company's chief test pilot Claes Smith at the controls. The first prototype was powered by a Swedish-built Nohab/Bristol Mercury XII of 880 hp, the second by a Pratt & Whitney R-1830 Twin Wasp of 1,065 hp.

The aircraft had very sleek lines and many advanced design features, including flush-riveting for low drag. For maximum strength the centre section

Two Saab B 17B bombers. *(Flygvapnet/F 7)*

A B 17A, with SFA-built Twin Wasp engine, fitted with four early rocket-launchers. *(Flygvapnet)*

was designed without any cut-outs for the undercarriage. The rearward-retracting main undercarriage units and their covering doors were intended for use as air-brakes during dive-bombing, and the tailwheel was also retractable. The aircraft was also almost unique in having a retractable ski undercarriage which actually produced less drag than the wheel type. Even the water-based version which was normally equipped with floats was converted to have skis in the winter. The floats were of the Edo type manufactured under licence by Hägglund & Söner at Örnsköldsvik in Northern Sweden.

The fuselage, which had a roomy, high-visibility cabin for the pilot and observer/navigator/rear-gunner, contained an internal bomb-bay.

The Saab 17 carried up to 700 kg of bombs of various sizes from 50 up to 500 kg. In the internal bomb-bay a 250 kg bomb (or eight 50 kg) could be

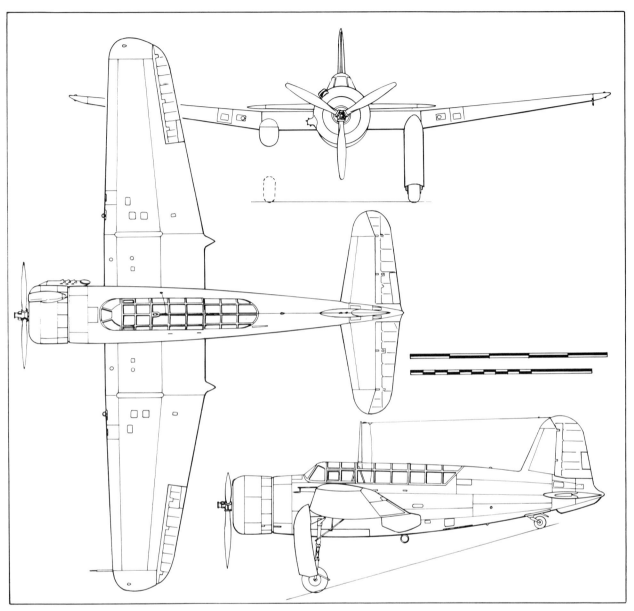

Saab B 17B

carried. The armament comprised two 8 mm machine-guns in the wings and one flexible 8 mm machine-gun at the rear seat. An early version of the aircraft used for classic dive-bombing tactics was equipped with a special 'fork', lowering the external carried 500 kg bomb free of the propeller arc.

During the production life of the aircraft the Saab BT-2 'toss' bombsight became available, making dive-bombing obsolete. This also made the use of the undercarriage doors as air-brakes unnecessary.

The reconnaissance version carried an N-2 camera in the fuselage.

The development and production of the Saab 17 was complicated by the problems of engine availability. Initially, the aircraft was planned for the Pratt & Whitney Twin Wasp. This version was designated B 17A in the Air Force. The Twin Wasp engine however did not become available until 1943–44, and therefore the first version of the aeroplane to go into production was the B/S 17B powered by a Swedish SFA-built Bristol Mercury XXIV of 980 hp. This engine also powered the seaplane version which was designated S 17BS. In 1941 the Air Force was able to procure the Italian Piaggio P XIbis RC 40 of 1,040 hp which powered the B17C version. The Wasp-powered B 17A thus became the last version to go into service. The engines in the B 17A and B had Hamilton Standard variable-pitch propellers built under licence in Sweden by Svenska Flygmotor. The Piaggio engines drove Piaggio P 1001 propellers.

In addition to the three prototypes,

The B 17 was the first aircraft to be equipped with the revolutionary Saab 'toss' bombsight. *(Flygvapnet)*

The undercarriage fairings of the B 17 could be used as air-brakes during dive-bombing. *(Flygvapnet)*

the Swedish Air Force ordered a total of 322 B 17/S 17s in four batches during the period 27 December, 1940, to 1 September, 1942. The first production aircraft flew on 1 December, 1941, and the last delivery took place on 16 September, 1944.

The type was manufactured both in Linköping and at Trollhättan. In fact, only 55 of the 322 production aircraft were completely built at Linköping.

The production of the four versions was split as follows: 132 B 17A; 116 B/S 17B (38 were delivered as seaplanes under the S 17BS designation); and 77 B 17C.

The Air Force career of the Saab 17, which began in early 1942, was very distinguished and six light-bomber and reconnaissance Wings (F 2, F 3, F 4, F 6, F 7, and F 12) were equipped with the aircraft. It was retired as a combat aircraft in 1948.

In the final phase of the Second World War it was feared that the German troops in Denmark (and Norway) would not obey Germany's order for total surrender. In Sweden the Danish Brigade, first organized in 1943, also included a number of Danish Air Force officers who in 1944 had been trained in the use of the Saab 17. Fifteen B 17Cs were actually allocated to the Brigade and were ready for deployment to Denmark and carrying Danish colours at the Swedish Air Force F 7 Wing at Såtenäs, but the order to fly to Denmark never came from the Danish Government.

In the period 1947-53 the Ethiopian Air Force, which had been organized by Swedish officers after the war at the request of Emperor Haile Selassie, eventually procured a total of 47 Saab 17As in three batches. Responsible for the organization of the Ethiopian Air Force was the Swedish Colonel Count Carl Gustaf von Rosen. The Saab 17s proved ideal for the rugged conditions in Ethiopia. In the late 1950s a number of Fairey Firefly attack aircraft were acquired from Canada in order to modernize the Air Force but these aircraft finished their service in Ethiopia well before the Saab 17s, which were still operating in squadron strength in 1960. Even in the 1970s some Saab 17s were operating in Ethiopia after more than 25 years of service in that country. The Ethiopian Saab 17s had their main base at Asmara, 2,300 m above sea level. The Saab 17s endurance of more than 4 hours was vital in that part of the world. Ethiopia, with its many high mountains and few airfields, covers twice as large an area as Sweden which itself is as big as the Federal Republic of Germany, Belgium, The Netherlands, Switzerland and Austria combined.

Starting in 1951, the Air Board released a number of Saab 17As to serve as civil registered target-towing aircraft for the Swedish armed forces. The aircraft were still owned by the Air Board but operated by private companies, Svensk Flygtjänst and AVIA, the latter company based on the island of Gotland. Eventually, a total of 20 Saab 17s were on the Swedish civil aircraft register, most of these serving with Svensk Flygtjänst. Two S 17BS seaplanes were also in civil use during 1949-51 owned by Ostermans Aero AB. One ex-Flygtjänst target-towing 17A was sold to Austria in 1957 and two years later two similar aircraft went to the Finnish Air Force.

A B 17A carrying a 500 kg bomb externally. *(Flygvapnet)*

A B 17C with Piaggio engine. *(Saab)*

In the period 1947-1953 the Imperial Ethiopian Air Force acquired a total of forty-six Saab 17As. They were still operated in squadron strength in 1960.

Because of its clean lines the Saab 17 was sometimes mistaken for a fighter by foreign intruders. *(Flygvapnet)*

S 17BS was the designation of the float-equipped version of the B 17. *(Flygvapnet)*

B 17A

Span 13.7 m (45 ft 1 in); length 9.8 m (32 ft 2 in); height 4.0 m (13 ft 1 in); wing area 28.5 sq m (307 sq ft). Empty weight 2,600 kg (5,732 lb); loaded weight 3,970 kg (8,752 lb). Maximum speed 435 km/h (270 mph); cruising speed 390 km/h (242 mph); landing speed 125 km/h (78 mph); initial rate of climb 10 m/sec (1,968 ft/min); ceiling 8,700 m (28,500 ft); range 1,800 km (1,120 miles).

B 17B

Span 13.7 m (45 ft 1 in); length 9.8 m (32 ft 2 in); height 4.0 m (13 ft 1 in); wing area 28.5 sq m (307 sq ft). Empty weight 2,635 kg (5,800 lb); loaded weight 3,835 kg (8,450 lb). Maximum speed 395 km/h (245 mph); cruising speed 375 km/h (233 mph); landing speed 125 km/h (78 mph); initial rate of climb 9 m/sec (1,770 ft/min); ceiling 8,000 m (26,250 ft); range 1,400 km (870 miles).

S 17BS

Span 13.7 m (45 ft 1 in); length 10.0 m (32 ft 10 in); height 4.8 m (15 ft 9 in); wing area 28.5 sq m (307 sq ft). Empty weight 2,700 kg (5,950 lb); loaded weight 3,800 kg (8,370 lb). Maximum speed 330 km/h (205 mph); cruising speed 315 km/h (196 mph); landing speed 125 km/h (78 mph); ceiling 6,800 m (22,300 ft); range 2,000 km (1,245 miles).

B 17C

Span 13.7 m (45 ft 1 in); length 10.0 m (32 ft 10 in); height 4.15 m (13 ft 7 in); wing area 28.5 sq m (307 sq ft). Empty weight 2,680 kg (5,900 lb); loaded weight 3,870 kg (8,525 lb). Maximum speed 435 km/h (270 mph); cruising speed 370 km/h (230 mph); landing speed 125 km/h (78 mph); ceiling 9,800 m (32,150 ft); range 1,700 km (1,060 miles).

Saab 17 production serials

L-10 (prototypes): 17001, 17002
B 17A: (SFA/Twin Wasp-powered bomber version); 17006, 17238-17368
B 17B: (SFA/Mercury-powered bomber version): 17003-17005, 17007-17016, 17101, 17105-17115, 17151-17164, 17187-17202
S 17BL: (land-based reconnaissance version): 17103, 17131-17150
S 17BS: (water-based reconnaissance version): 17104, 17116-17130, 17165-17186
B 17B: (Piaggio-powered bomber version): 17017, 17057, 17102, 17203-17237

Civil registered Saab 17s

Saab 17A

17239	SE-BYH	To Flygvapen Museum
17249	SE-BUD	
17251	SE-BYG	
17256	SE-BYE	
17267	SE-BZH	
17268	SE-BRN	
17284	SE-BRR	
17296	SE-BPP	
17308	SE-BPR	
17313	SE-BUM	
17318	SE-BWA	
17320	SE-BWC	To Finland as SH-2
17334	SE-BUL	
17336	SE-BYK	
17339	SE-BYF	To Austria
17355	SE-BRO	To Finland as SH-1
17356	SE-BUN	
17358	SE-BUK	
17364	SE-BUH	

Saab 17BS

17174	SE-APC	Ostermans Aero AB
17185	SE-BFA	

The B 18A served as a bomber only for a brief period and was converted for strategic reconnaissance as the S 18A. It was the first Swedish aircraft with radar. *(Saab)*

Saab 18

In late 1938 the Swedish Air Force invited the aircraft industry to bid for a twin-engined bomber suitable for dive-bombing and strategic reconnaissance. Three companies, Saab, ASJA and Götaverken, responded. The aircraft was to have a crew of three, have an internal bomb-bay and be capable of dropping bombs from dive angles up to 85 degrees. Initially only the 1,065 hp Pratt & Whitney R-1830 Twin Wasp would be available but later on more powerful engines such as the Bristol Taurus of 1,215 hp were envisaged. With the latter engine, a top speed of 550 km/h (342 mph) was required.

Of the three proposals, Götaverken's (project manager Bo Lundberg) was the most advanced but was also the most expensive. Saab's proposal (project manager Alfred Gassner) was not accepted. The ASJA project (project manager Bror Bjurströmer) was acceptable but needed some modifications.

Götaverken's proposal fell through because the company was not able to provide the production facilities which were an Air Force condition for awarding a prototype contract. Eventually, Götaverken (and its successor AB Flygplanverken with Bo Lundberg as general manager) had to pull out of aircraft manufacturing altogether.

Early in 1939, Saab and ASJA merged and formed the 'new' Saab with headquarters at Linköping. The L-11 project orginally submitted by ASJA could eventually be modified to meet the Air Force requirements and in November 1939 a first prototype was ordered and designated Saab 18A. In February 1940 a second prototype was ordered.

The development of the aircraft was difficult and time-consuming, not least as a result of considerably modified Air Force requirements such as a completely new bomb installation, self-sealing fuel tanks, extra fuel tank for the bomb-bay, armour protection for the crew, a fixed gun installation in the nose, and a modified nose with the bomb aimer and his bombsight moved slightly to starboard to enhance the pilot's view. As the originally meagre Air Force budgets were increased as the war continued, offering possibilities of longer production runs, the interest of the industry and its engineering work force was greatly stimulated. In 1939, however, it was still difficult for the Air Force to judge the perfor-

The B 18B represented a major improvement in Swedish attack capability against sea invasion. *(Flygvapnet/F 7)*

The T 18B was one of the fastest twin-engined piston-engined bombers.

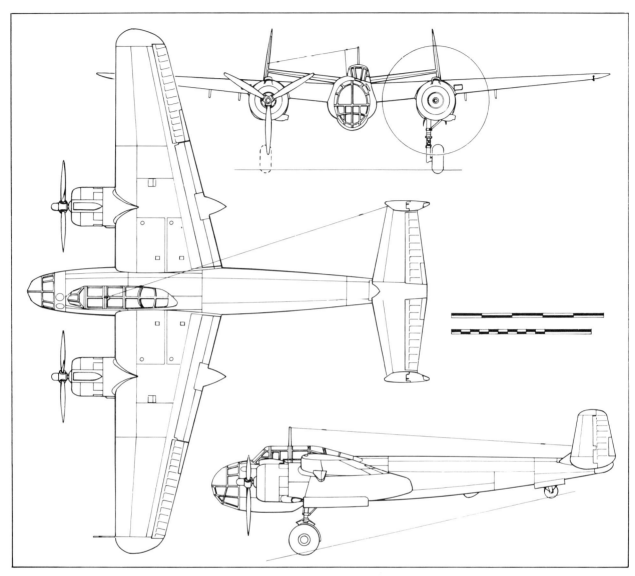

Saab B 18A

mance of the new Swedish aircraft industry in general and Saab in particular.

The first Saab 18 prototype flew for the first time on 19 June, 1942, and the second soon after, both powered by Twin Wasp engines. On 31 July, 1942, the Air Force placed an initial order for 62 aircraft designated B 18A.

Deliveries started in March 1944 and continued until December 1945. All aircraft went to the Air Force Wing at Västerås (F 1) where they replaced the venerable Junkers Ju 86K (B 3). Service introduction was not without its problems and in late 1944 all engine mountings had to be modified, disrupting the conversion training. By February 1945, however, all three F 1 squadrons had converted to the B 18A.

The B 18A represented a major improvement of the striking power of the bomber squadrons, and air-defence exercises showed that the bombers were no longer such easy prey to the fighters as they had been during the B 3 era. Repeated attacks were no longer possible, the speed difference being too small.

The B 18A was able to carry a total of 1,400 kg of bombs, 40 percent more than the B 3. The internal bomb-bay could accommodate two 500 kg bombs or three 250 kg bombs, and alternatively, ten 50 kg bombs. Under the wings eight 50 kg bombs or flare bombs could be carried. The armament comprised one fixed forward-firing and two flexible 13.2 mm machine-guns.

The B 18A did not serve the F 1 Wing very long. Already during 1946 the first few aircraft were transferred to the strategic reconnaissance Wing (F 11) at Nyköping and converted to their new role as the S 18A. The modification progamme, which was mainly done by the central workshops at

All Saab 18 versions initially carried a crew of three. This is the entrance door on the T 18B version. *(Saab)*

The Saab 18 in three different versions. OPPOSITE TOP: The Saab 18A (B 18A) which made its first flight in 1942 powered by two SFA/Twin Wasp engines; OPPOSITE BOTTOM: The Saab 18B (B 18B) first flew in 1944 powered by Daimler-Benz DB 605Bs; and ABOVE: The T 18B was originally designed to carry torpedoes but later modified to carry heavy cannon armament (one 57 mm and two 20 mm) and rockets. *(Saab and Flygvapnet)*

Västerås (CVV), included installation of an SKa 5 (Hasselblad) vertical camera and an SKa 10 camera in the glazed nose. At the end of 1949, CVV also began installation of a surveillance radar of the United States type AN/-APS-4 (PS-18A in the Swedish Air Force) as well as a radio altimeter, PH-10/A. The S 18A thus became the first Swedish Air Force aircraft to carry radar. The equipment was installed in a pod beneath the nose and the radar had a range of approximately 100 nautical miles. For night photography the aircraft was equipped with the SKa 13 camera which produced excellent results in combination with flares.

The S 18As were split about equally between the F 11 Wing and the F 3 Wing at Malmslätt near Linköping. Later on, the aircraft also served with the F 21 Wing at Luleå in the north.

The last S 18A was not retired until 22 May, 1959, when it was finally replaced after 15 years of Air Force service by the Saab 32 Lansen transonic jet aircraft. The S 18A survived the S 31 (Spitfire Mk. 19) also used by the Air Force for high-altitude missions and served alongside the Saab 29 jet reconnaissance aircraft. Two S 18As were also on the civil register, being operated by Airborne Mapping Ltd in 1957-60. They were SE-CFL (18131) and SE-CFO (18157).

The more powerful version of the Saab 18, planned from the start, made its first flight on 10 June, 1944, and was powered by two 1,475 hp Daimler-Benz DB 605Bs for which a manufacturing licence had finally been obtained. Designated B 18B, this version was much faster than its predecessor, having a top speed of 570 km/h (354 mph). A later version, the T 18B, was even faster with a top speed of 595 km/h (370 mph) making it one of the fastest bombers of the Second World War.

The B 18B was really the first true attack aircraft in the Swedish Air Force since the bomb had ceased to be the main weapon for this kind of aircraft for which completely new tactics were now developed. A new version of the Saab 'toss' bombsight (BT9) was installed in the B 18B allowing shallow dive bombing. The B 18B was also the first Swedish aircraft to carry rockets, and these were normally carried under the wings and below the fuselage nose (12 in total) of various calibres. For this role a new gunsight was installed and the crew reduced to two, the lower flexible gun being removed.

At the beginning of 1949 the B 18B was subjected to a modification which was probably unique in the world. The aircraft in service were returned to Saab for installation of two ejector seats of a type which in the meantime had been developed for a Saab fighter.

A total of 120 B 18Bs were delivered between October 1945 and February 1949. They served with the bomber

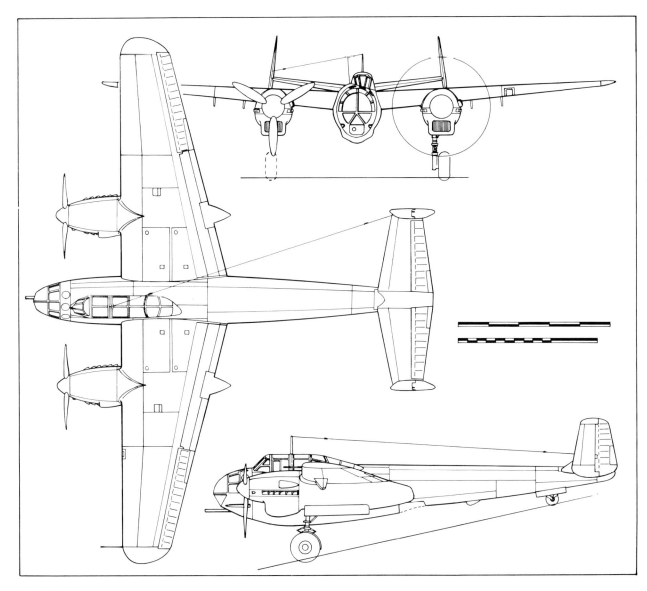

Saab T 18B

(later attack) Wings at Halmstad (F 14), Västerås (F 1) and Såtenäs (F 7). The last B 18B was retired in 1958.

A third Saab 18 version designated T 18B was also developed. It was originally adapted for dropping torpedoes and mines but this requirement was later scrapped for various reasons. The 45 cm Norwegian torpedoes suitable for high speeds and intended for use by the T 18B did not become available in time; instead the aircraft was armed with two built-in 20 mm cannon. The aircraft could also be equipped (in the torpedo bay) with a 57 mm cannon of Bofors manufacture (L/50). The cannon had a weight of 735 kg (1,620 lb) and measured 5.3 m (17 ft 3 in) in length. 40 rounds were carried and the rate of fire was 2 per second. The cannon could be installed in only two hours. The precision was excellent (at distances of up to 2 km). The recoil force was nearly 6 tonnes but surprisingly did not significantly affect the aircraft's characteristics due to a successful recoil-brake installation. The T 18B was also used for testing experimental anti-ship missiles intended for the advent of new types of attack aircraft. The Rb 302 missile tested in the T 18B led to the development of the successful Rb 04, variants of which are still being used.

Sixty-two T 18Bs were delivered to the F 17 Wing at Ronneby between June 1947 and December 1948. In the summer of 1957 its replacement with jet aircraft began and in January 1958 the last T 18B was retired.

A total of 244 Saab 18 aircraft were delivered.

In a unique conversion, all B 18Bs and T 18Bs were retrofitted with Saab ejector seats for both pilots and navigator/gunner. The third crew member was eliminated when the bomb ceased to be the main armament of bomber (now attack) squadrons. *(Saab)*

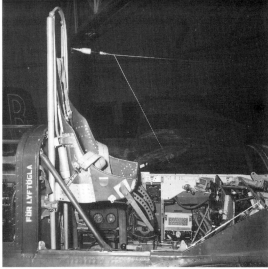

B 18A/S 18A

Span 17.04 m (55 ft 10 in); length 13.23 m (43 ft 5 in); height 4.35 m (14 ft 3 in); wing area 43.8 sq m (470 sq ft). Empty weight 5,484 kg (12,080 lb); loaded weight 8,693 kg (19,165 lb). Maximum speed 465 km/h (289 mph); cruising speed 415 km/h (258 mph); landing speed 135 km/h (84 mph); ceiling 8,000 m (26,250 ft); range 2,200 km (1,370 miles).

B 18B

Span 17.04 m (55 ft 10 in); length 13.23 m (43 ft 5 in); height 4.35 m (14 ft 3 in); wing area 43.8 sq m (470 sq ft). Empty weight 6,093 kg (13,433 lb); loaded weight 8,793 kg (19,385 lb). Maximum speed 570 km/h (354 mph); cruising speed 480 km/h (298 mph); landing speed 125 km/h (78 mph); ceiling 9,800 m (32,150 ft); range 2,600 km (1,620 miles).

T 18B

Span 17.04 m (55 ft 10 in); length 13.23 m (43 ft 5 in); height 4.35 m (14 ft 3 in); wing area 43.8 sq m (470 sq ft). Empty weight 6,183 kg (13,630 lb); loaded weight 9,272 kg (20,420 lb). Maximum speed 595 km/h (370 mph); cruising speed 480 km/h (298 mph); landing speed 130 km/h (81 mph); ceiling 9,300 m (30,500 ft); range 2,600 km (1,620 miles).

Saab 18 production serials

Prototypes: 18001, 18002 (18001 later re-engined with DB 605B engines as prototype for B 18B).
B 18A/S 18A: 18101–18162
B 18B: 18163–18282
T 18B: 18283–18343

To test the geometry and characteristics of the Saab 21A nosewheel undercarriage, an Sk 14 (North American NA-16-4M) was modified. *(Saab)*

One of the first inflight photographs released in early 1945 of the Saab 21A pusher-propeller fighter. The unusual configuration enabled a strong concentration of armament to be made in the nose. *(R Wall)*

Saab 21

From November 1940 in the Basic Agreement between the Air Board/(KFF)* and the industry, development of a fighter aircraft was included. From early 1941, alternative solutions were discussed and on 1 April, the L-21 project was presented by Saab for the Air Force. The L-21 was a refined version of the very preliminary L-13 radial-engined project of late 1939. The powerplant was a Swedish-built version of the Daimler-Benz DB 605B of 1,475 hp. The L-21 was a very unconventional design featuring a rear-mounted engine and pusher propeller and with the tail unit supported by twin booms. These also housed the main under-

* Kungliga Flygförvaltningen was the full name of the then Swedish Air Force Board.

carriage and machine-guns (13.2 mm later replaced by 12.7 mm guns) were sited in the forward part. Otherwise the armament was concentrated in the fuselage nose which housed one 20 mm cannon and two more 13.2 mm machine-guns. The aircraft was the first in Sweden to have a nosewheeel undercarriage. Project manager was Frid Wänström.

There were initially quite a few technical uncertainties in the project, such as: 1) How would the engine cooling on the ground be arranged as there was no propeller slipstream? 2) Would the gun-powder-propelled ejector seat provide the pilot with sufficient clearance of the propeller in the event of a baleout? 3) How would the nosewheel undercarriage function on Sweden's grass airfields? 4) What would the effect be on control characteristics on the ground with the rudders outside the

propeller slipsteam? These were some of the major questions raised by the Air Board. The uncertainties prevailed for some time and in October 1941 the Air Board stopped the initial project work ordered in April; instead Saab was instructed to start project work on a more conventional back-up design, the L-23, in general layout similar to that of the North American P-51 Mustang. Eventually, Saab managed to convince the Air Board of the advantages of the L-21 solution. Superior pilots visibility, armament concentration, and safeguard against ground loops were some of the justifications. In November, Saab received a final go-ahead for the L-21. A mock-up was approved on 8 July, 1941.

On 5 December, 1941, the project was handed over to the design office in Linköping for materialization. Chief of the design office was A. J. Andersson.

An unusual view of a formation of J 21As. (Flygvapnet/F 15)

The technical problems mentioned were eventually solved. Ground cooling was arranged by installation in the inner wing of two cooling fans driven by the engine via a mechanical gearbox. The fans rotated at 13,500 rpm when the engine rpm was 1,800. The fans were automatically disconnected in flight. The ejector seat, the first of its kind at that time, was a major development task. Initial ground tests with a wooden dummy were followed on 22 February, 1944, by flight tests using a Saab 17. The tests were very promising and the first J 21 prototype was equipped with an ejector seat from the very first flight on 30 July, 1943, although some further testing remained to be done.

The new tricycle undercarriage required another major development effort. To validate estimates made on the geometry, nosewheel shimmy etc, a test rig with the same mass as the aircraft was built and towed behind a car under various conditions. The next step was the conversion of an Sk 14 (North American NA-16-4M) trainer to a tricycle configuration. These tests were successful and removed all remaining doubts.

The nosewheel undercarriage of the Saab 21A was another novel feature for a fighter in the early 1940s. *(Saab)*

To achieve the high speed required for the new fighter, a new wing profile had to be developed in order to reduce drag and to maintain laminar flow over as much of the wing area as possible without losing too much lift. The wing profile which was developed at Saab and extensively tested in the wind tunnels at the Aeronautical Research Establishment (FFA) and the Royal Institute of Technology (KTH), both in Stockholm, proved very successful, while the interference drag of the

The Saab 21 proved to be an excellent weapons platform and a special attack version, the A 21A-3, could carry up to 800 kg of bombs on central and underwing racks. *(Flygvapnet/F 7)*

Saab J 21A

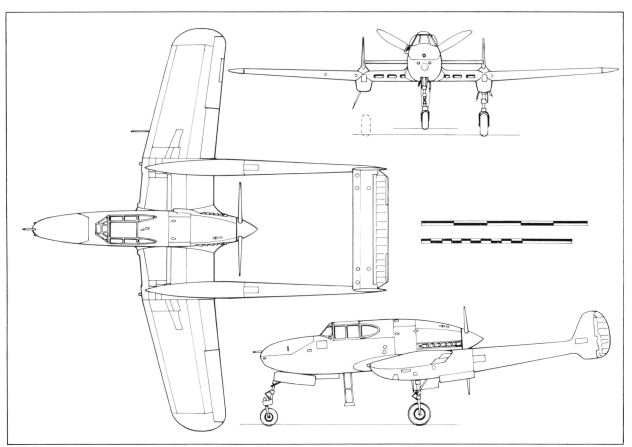

tail booms was considerably less than that of conventional engine nacelles. The flight testing confirmed the early estimates by giving a top speed 25 km/h (15½ mph) higher than the guaranteed speed. Interestingly, the Saab 21 wing profile proved to be very similar to that chosen for the British Hawker Tempest fighter.

The Saab 21 was indeed a very unconventional aircraft by any standard but the first flight in 1943 could well have been a disaster. The aerodynamicists had for some reason come to the conclusion that the take off should take place with full flap. Following wind-tunnel tests this should have given the shortest possible take-off distance, according to some experts. The Saab chief test pilot Claes Smith voiced some doubts but eventually the Air Force test pilots consulted also gave in. At the expected point of rotation nothing happened and it was too late to stop.

Saab's grass aerodrome was very short and surrounded by a road and a low fence. Instead of stopping, the pilot built up speed as much as possible. The aircraft touched the road and tore down the fence but was airborne. The undercarriage could, however, be retracted and the flight characteristics checked out. For the landing the larger airfield at Malmslätt was chosen. During the smooth landing, however, the pilot discovered that there was no wheel-braking due to damage caused at take off. The pilot suddenly recalled that he had an anti-spin parachute installed and released it, thus inventing the brake chute! At this moment the main undercarriage started to collapse causing the propeller tips to function as a very effective brake and collision with the surrounding forest could be avoided. Damage proved to be limited and the aircraft was repaired quickly. Flight testing was now moved to the much larger airfield at F 7 at Såtenäs in western Sweden where the flight testing eventually became a less exciting routine.

Once the teething troubles, mainly ground cooling, were overcome – the Saab 21 became a robust and well-liked aircraft. It was easy to fly and the nose-wheel undercarriage greatly facilitated taxi-ing. Pilots liked the exceptionally small turning radius, good stall characteristics and general stability. The positioning of the pilot immediately behind the guns gave excellent firing precision, the three nose guns giving a very concentrated burst of fire – at 450 m (1,500 ft) range less than two feet wide. Boom weapons were harmonized at 200 m (720 ft) and spread was 2 m (6 ft 6 in). Since in 1945 the Air Force was able to acquire the American P-51D Mustang (J 26 in Sweden) there were plenty of opportunities to compare the two fighters. Due to a more

To enable the pilot to clear the propeller of the Saab 21A in the event of an inflight emergency, Saab developed an ejector seat, one of the first in the world. This film strip is from a test using a dummy ejected from a Saab 17 in February 1944. *(Saab)*

The instrument panel of the J 21A. The front panel was of 100 mm thick armoured glass. *(Saab)*

The J 21A's bombs soon gave way to rockets. A normal load was eight 15 cm rockets under the outer wings and two 18 cm rockets under the centre wing, a quite formidable fire power for its time. *(Flygvapnet/F 6)*

powerful compressor-equipped engine and more modern aerodynamics, the Mustang not unexpectedly proved superior in turning performance and climb, but only marginally. Often the superior pilot visibility and fire-power concentration of the J 21 could be decisive in mock combat when combined with individual pilot skill.

Special tactics adapted to the qualities of the aircraft were developed and the tight turning and favourable stall characteristics could be used to advantage in a dog fight. An effective manoeuvre was to climb vertically and wait until the pursuing fighter stalled. After that the J 21 pilot could wheel over and dive on him.

The very fact that the J 21 proved to be an excellent gun platform, with good low-level characteristics and pilot visibility, made it natural for it to be used as an attack aircraft. A special attack version was also developed in addition

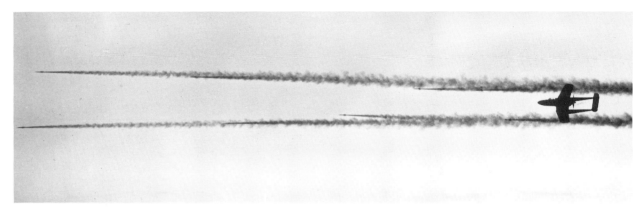

This Saab J 21A has just launched its wing-mounted rockets.

In 1946 the J 21As were sometimes manhandled out of their underground shelters. *(Flygvapnet/F 12)*

to two fighter versions.

The first Air Force unit to receive the J 21 was the F 8 fighter Wing at Barkarby near Stockholm, with the intention of using them for Service trials and conversion training. The first 15 aircraft received were later allocated to the F 9 fighter Wing at Gothenburg where the J 21 was operated until replaced by the J 28B Vampire in 1950. The next fighter Wing to receive the aircraft was F 15 at Söderhamn which used it until 1953. The F 12 Wing at Lalmar converted from Saab 17s to J 21 fighters in 1947. The two Wings, F 6 at Karsborg and F 7 at Såtenäs, received the attack version of the aircraft in 1947-49.

Altogether, a total of 302 Saab 21s were delivered, including 59 J 21A-1s (the basic fighter version), 124 J 21A-2s (fighter version with modified inner wing flaps and cooling) and 119 A 21A-3s with attack armament. The two fighter versions differed in that the J 21A-2 carried a 20 mm Bofors cannon instead of the Hispano gun in the A-1 as well as a K-14 gunsight.

The attack version could carry a 600 kg (1,320 lb), 500 kg (1,100 lb), and 250 kg (550 lb) bomb on the inner-wing centre pylon and in addition four 50 kg (110 lb) bombs under each outer wing. As an alternative to bombs, eight 14.5 cm and two 18 cm rockets could be carried. A Saab BT 9 toss-bomb computer was also installed. The gun armament was unchanged compared to the A-2 fighter version. It also carried 400 litres of extra fuel in each wingtip tank compared to 160 litres in each tank in the A-1/A-2 versions.

J 21A-1/A-2

Span 11.60 m (38 ft); length 10.45 m (34 ft 3 in); height 4.0 m (13 ft 1 in); wing area 22.2 sq m (239 sq ft). Empty weight 3,250 kg (7,165 lb); loaded weight 4,150 kg (9,149 lb). Maximum speed 640 km/h (398 mph); cruising speed 495 km/h (308 mph); landing speed 145 km/h (90 mph); initial rate of climb 15 m/sec (2,950 ft/min); ceiling 11,000 m (36,100 ft); range 750 km (466 miles).

A 21A-3

Span 11.60 m (38 ft); length 10.45 m (34 ft 3 in); height 4.0 m (13 ft 1 in); wing area 22.2 sq m (239 sq ft). Empty weight 3,250 kg (7,165 lb); loaded weight 4,413 kg (9,729 lb). Maximum speed 640 km/h (398 mph); cruising speed 495 km/h (308 mph); landing speed 145 km/h (90 mph); initial rate of climb 15 m/sec (2,950 ft/min); ceiling 11,000 m (36,100 ft); range 1,500 km (930 miles).

Saab 21 production serials

Prototypes: 21001, 21002, 21003
J 21A-1: 21101, 21103-21115, 21117, 21118, 21120, 21122, 21124-21159
J 21A-2: 21160-21183
A 21A-3: (Attack version): 21102, 21343-21402

The Saab R 21R only equipped one fighter wing, F 10 at Ängelholm. *(Saab)*

Saab 21R

In order to provide early experience of jet aircraft, early in 1945 Saab proposed an inexpensive conversion of the existing Saab 21A piston-engined fighter. To power the Saab 21R, as the new aircraft was designated, the British de Havilland Goblin engine of 1,360 kp (2,996 lb) static thrust, was chosen. The centrifugal-compressor Goblin was of much greater diameter than the DB 605B twelve-cylinder inline engine it replaced but actually the larger diameter facilitated the air-intake design. A condition for the project was that the conversion should only include such changes as were absolutely necessary in view of the engine installation and the higher speed. Other changes included the tail unit which had to be considerably modified, the elevator being raised above the jet blast. It was also structurally stiffened to eliminate the risk of tail flutter which in fact had been an early problem with the original Saab 21. Originally it was hoped that some 80 percent of the 21A design would remain in the 21R; eventually, only 50 percent remained.

To improve the aerodynamic configuration of the wing, the inner wing chord was somewhat extended and sharpened. Furthermore, the windscreen was streamlined, and a large number of structural inspection doors were strengthened.

The higher fuel consumption required extra fuel space, which was made available by elimination of the bulky oil and coolant radiators. The fuel system was redesigned to match

A squadron of Saab 21Rs being towed from their shelters. *(Flygvapnet/F 10)*

The Saab 21R *(left)* was the first fighter converted from piston-engine power (J 21A *right*) to jet propulsion. It had an SFA-built de Havilland Goblin. *(Saab)*

The Saab 21R was later used by two attack Wings and was also equipped with an external pod *(lower photograph)* housing as many as eight machine-guns, giving a total of 13 guns. *(Flygvapnet/F 7 and F 17)*

the new propulsion system, and an oil heating system was added since the British engine was not adapted to Arctic conditions.

The absence of propeller slipstream acting on the tail surfaces to assist in lifting the nosewheel had to be compensated. For this reason the main undercarriage wheel centres were re-positioned about 200 mm (8 in) forward and 300 mm (12 in) higher to achieve a favourable rotating balance. In the original design, the air-brakes were positioned in the undersurface of the wing centre section in lieu of the old radiator flaps. This location, however, caused marked trim changes at high speed which made it necessary to move the air-bakes to the trailing edge of the outer wing where they worked on the principle of double split-flaps.

Other changes included installation of a new Dowty Constant Pressure hydraulic pump which operated the undercarriage, flaps, air-brakes and wheel brakes. In addition, there was a compressed-air system for lowering the wheels and ground braking.

The cockpit layout was completely different to the 21A's, with new instruments, improved gunsights, a 'demand' type oxygen system, and new radio equipment.

The first of four prototypes made its first flight on 10 March, 1947, with Åke Sundén at the controls. The flight took place only about one year after the start of design work. Ragnar Härdmark was the project manager.

A major problem which remained with the aircraft was the low critical Mach number of the wing, which could easily be exceeded and resulted in a nose-down tendency. Climb and dive performance were also unacceptable for the fighter role, particularly above 6,000m (19,700 ft). Tight turning capability and superb take-off and landing characteristics even during adverse conditions were, however, merits appreciated by the pilots.

The fighter combat tactics developed for the 21R were based upon German Messerschmitt Me 262 experience, with hit-and-run and head-on attacks followed by a quick pursuit curve.

The only Air Force fighter Wing to be equipped with the J 21R was F 10 at Ängelholm which received its first aircraft in August 1949. Owing to its performance limitations, the 21R's service life as a fighter aircraft was brief and already in the late spring of 1950 the transfer of the F 10 aircraft to the F 7 attack Wing began. In the meantime, the planned series of 120 aircraft had been reduced to 60 in favour of a production acceleration of the new swept-wing J 29 fighter already undergoing flight testing at Saab.

In the attack role the A 21R (the new designation), which introduced jet aircraft to the attack units, won many friends among the pilots. New tactics and weapons were introduced, stressing the importance of low-level missions. After serving with F 7 until 1954 the remaining A 21Rs were transferred

Åke Sundén (LEFT) made the initial flight with the first Swedish jet fighter in March 1947; Ragnar Härdmark (RIGHT) was the J 21R project manager. *(Saab)*

Saab J 21R

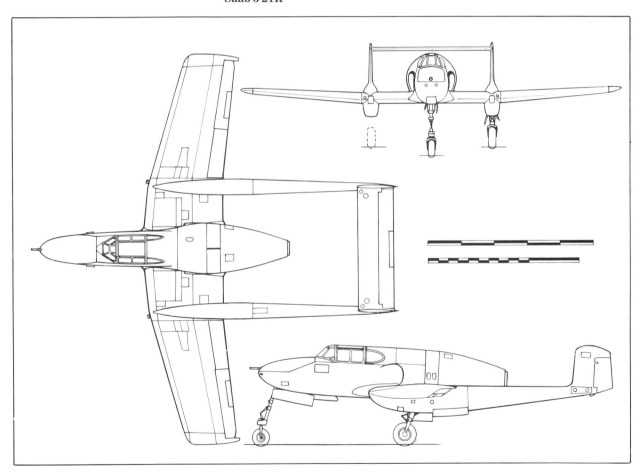

In August 1951 it was still permissible to demonstrate to the public rocket firing inside the airfield perimeter, in this case at Malmslätt near Linköping. *(Östgöta Correspondenten)*

to the F 17 attack Wing at Ronneby. They were retired in 1956. The main weapons of the attack version were light and heavy rockets (15 and 18 cm calibre) of which up to 10 could be carried in various combinations in addition to the fixed armament of five guns (one 20 and four 13.2 mm). A special weapon alternative developed for the A 21R was an external gun pod slung under the inner wing and containing eight 8 mm machine-guns. The simultaneous use of all thirteen guns was quite a sensation even for the pilot, because of the recoil forces!

As a weapons platform the aircraft was extremely stable and easy to aim against the target. The greatest weakness was the performance limitations with heavy rocket loads which caused the A 21R squadrons to concentrate their training on very low level attacks to avoid radar detection.

Including prototypes, a total of 64 aircraft were delivered of which 34 were of the J 21RA version powered by the de Havilland Goblin 2 (RM 1) of 1,360 kp (2,966 lb) and 30 J 21RB powered by a Svenska Flygmotor-built Goblin 3 (RM 1A) of 1,500 kp (3,304 lb) static thrust. The latter version was delivered between July 1950 and January 1951.

The Saab 21R is not likely to go down in history as a major accomplishment by the Swedish aircraft industry but it was indeed a very useful introduction to jet air intake design, high-speed aerodynamics, and high-speed control systems – all of great value for later, more advanced aircraft.

The J 21R was delivered from 1949 until 1951. This production picture was taken in August 1949. *(Saab)*

J21RB

Span 11.37 m (37 ft 4 in); length 10.60 m (34 ft 9 in); height 2.90 m (9 ft 6 in); wing area 22.30 sq m (240 sq ft). Empty weight 3,112 kg (6,860 lb); loaded weight (normal) 5,033 kg (11,095 lb). Maximum speed 800 km/h (497 mph); cruising speed 700 km/h (435 mph); landing speed 155 km/h (96 mph); initial rate of climb 17 m/sec (3,345 ft/min); ceiling 12,500 m (41,100 ft); range 900 km (560 miles).

Saab J 21R Production serials

Prototypes (converted J 21A-1): 21116, 21119, 21121, 21123
J 21rA: 21403–21409, 21411–21433
J 21rB: 21410, 21434–21462

The Scania prototype over Östergötland, the home province of Saab. (Saab)

Saab 90 Scandia

As early as the end of 1943 the first discussions regarding civil aircraft started at Saab. The military workload could hardly be expected to continue unchanged after the end of hostilities and a complement would be needed if the industry were to be able to survive and develop. In this situation Saab contacted ABA Swedish Air Lines and at the beginning of 1944 a meeting took place between Saab's managing director Ragnar Wahrgren and ABA's managing director Capt Carl Florman. ABA saw an early need for new equipment and summarized its requirements for a twin-engined aircraft in a document handed over to Saab on 26 January, 1944. The aircraft was intended primarily to replace the Douglas DC-3s already in the ABA fleet. ABA also recommended selection of an established engine type such as the Pratt & Whitney R-1830 Twin Wasp. The importance of good low-speed characteristics was particularly stressed. A combined civil transport and bomber aircraft was also discussed with the Air Force but ABA strongly advised against such a 'violent compromise which would not meet either fundamental requirements. A combination of commercial and military transport/ambulance aircraft would naturally be possible', ABA concluded.

On 23 February, 1944, a preliminary project description was made available to ABA in which every effort had been made to incorporate the airline's requirements. ABA, after all, had 20 years of experience of scheduled airline operations. The project called for a short to medium-haul twin-engined low-wing all-metal aircraft seating 25–30 passengers plus cargo. The range would be about 1,000 km (620 miles) and take-off weight in the region of 11,600 kg (25,550 lb); the engines Pratt & Whitney R-1830s and the propellers of the Swiss Escher-Wyss type. At this point some market investigations were made in addition to the matching to ABA's requirements.

Even on the drawing board, however,

The prototype 90A-1 Scandia after its first flight on 16 November, 1946. (Saab)

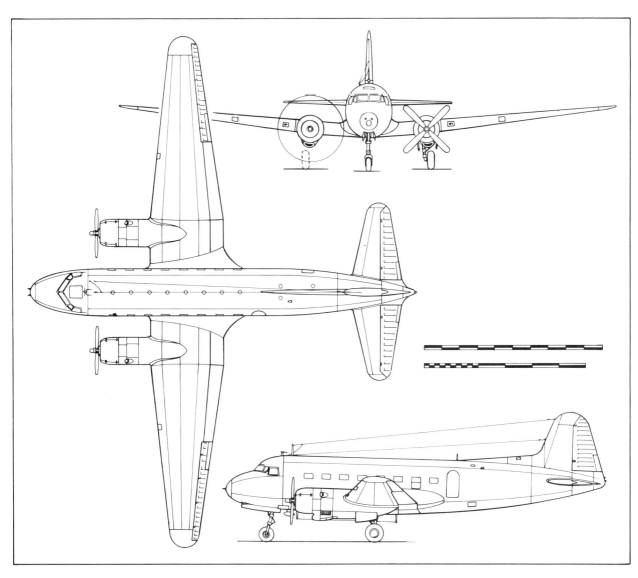

Saab 90 Scandia

it was found that the R-1830 was not powerful enough and it was subsequently replaced by the R-2000 of 1,450 hp. On 28 February, the project was submitted to Saab's Board of directors, who decided that development work should go ahead. For Saab the project would mean reduced dependence on military orders and a broader base for future development. For ABA there was a unique opportunity to have from the beginning an aircraft tailor-made to its requirements. It is important to stress, however, that ABA made no financial obligation to the project at this time.

The actual design work started in the spring of 1944. Technically responsible at Saab were Bror Bjurströmer, chief designer, and Tord Lidmalm, project manager. For ABA, Bo Hoffström, ABA technical project manager, and Karl Lignell, chief engineer, participated. During the autumn of 1944 a mock-up was completed as well as the assembly jigs. The original schedule had forecast a first flight in the summer of 1945, with production deliveries starting in 1947. During the work it became obvious that this schedule was too optimistic in view of the extensive new requirements to be met for a commercial aircraft. Another major problem was a six month long strike in the entire Swedish metal industry union.

The Saab 90 Scandia, as the project CT (Civil Transport) had now been named following an in-house competition among the Saab employees, was designed to meet the requirements of the US civil aviation authorities. These requirements, CAR 04, were the highest for a commercial aircraft of the Scandia's size and compliance with these was a necessity for international sales. Above all, the CAR rules called for much more extensive ground and flight testing than was ever required for military aircraft. There were still no orders for the Scandia but, at the end of

the final assembly of the prototype took place but further delays occurred and only in late October the roll-out of SE BCA, as the prototype was registered, became a fact. On 16 November the first flight took place with Claes Smith and Olle Hagermark at the controls. 'In contrast to the J 21 fighter, the Scandia behaved exactly as predicted on its first flight', Smith said after the nearly one-hour-long first flight. The flight testing went very much according to plan and as early as March 1947 the Board of Civil Aviation issued permission to undertake combined service trials and demonstration flights. These last began on 13 March when the Scandia went to Copenhagen and further out into Europe.

The continued testing led to some

The prototype Scandia after modification of the engine nacelles to a more conventional configuration. The engines were Pratt & Whitney R-2000s of 1,450 hp each *(Saab)*

PP-XE1, the third production Scandia, was intended for Swedish Air Lines as SE-BSC but was delivered to Aerovias Brasil and re-registered PP-SQE. Production Scandias had 1,825 hp Pratt & Whitney R-2180 engines and Hamilton Standard propellers. *(Saab)*

August 1945, ABA made a preliminary commitment for three unpressurized aircraft and 15 with a pressurized cabin.

Although the aircraft was not originally designed for a pressure cabin, projected development with a circular section fuselage featured one. Up till now the project had been worked on in relative secrecy but in December 1945 the Scandia was publicly announced. During the spring and summer of 1946 modifications, the most important of which was a modified engine cowling to improve cooling. Testing on unpaved aerodromes sometimes brought the propeller tips dangerously close to the ground – especially in the case of a flat

SAS's Scandia SE-BSB *Gardar Viking*, like all other SAS Scandias, was acquired by VASP of Brazil. This aeroplane, with 20,670 hr, flew more than any other Scandia.

In service, SAS's Scandia SE-BSL (*Folke Viking*)

tyre. For this reason the engines were raised some 150 mm (6 in) and the nosewheel leg 100 mm (4 in) while the engine cowling diameter was reduced by 75 mm (3 in) and given a more conventional shape with a separate carburettor intake below the engine. Different propeller types were also tested, including the Curtiss-Wright Electric, before the Hamilton Standard Hydromatic was selected as standard equipment.

During the continuing flight testing and demonstration flights in various parts of the world, it became evident that the prototype's engines were not powerful enough, the 1,450 hp Pratt & Whitney R-2000 being the same as those fitted to the Douglas DC-4. Consequently it was decided to equip the production version with a more powerful engine, the 1,825 hp R-2180-E1. Unlike the prototype, the Saab 90A-1, the 90A-2 production version was equipped with four-blade Hamilton Standard propellers instead of three-blade units.

The production Scandia could accommodate 24 or 32 passengers, the former on three-abreast seating, the latter four-abreast; the seats in the 24 passenger version were broad and reclining. The seats were all made by Saab. The passenger cabin had 16 windows, and there were also eight small round windows in the cabin ceiling which were designed in such a way that they incorporated built-in lamps to illuminate the cabin at night. At the rear of the cabin opposite the entrance door was a wardrobe and a small but very functional pantry. The lavatory also had a ceiling window. Behind the passenger cabin there was a cargo-hold with a volume of 6.4 cu m and below the cabin two more cargo holds, the front one 2.3 cu m capacity and the rear 2 cu m. The total cargo volume was 11 cu m, 1 cu m less than the Convair CV-240 but considerably more than the DC-3 (4.5 cu m) and Martin 2-0-2 (7.5 cu m). Structurally the aircraft was conventional. Each outer wing held two fuel tanks with a total volume of 2,960 litres (650 Imp gal). The undercarriage was completely retractable with interchangeable single-wheel main units, and a single nosewheel. The undercarriage retracted forwards to facilitate lowering in case of hydraulic failure. The cabin heater was a petrol-burning Stewart-Warner Southwind and three such heaters were also used for de-icing the wing leading edge and the vertical fin leading edge. The standard equipment also included a Sperry A-12 autopilot but the radio equipment differed with customers' requirements.

The marketing effort was intensive and included several extensive demonstration tours in Europe, Africa, and later the United States and Latin America.

On 20 April, 1948, ABA finally signed a firm contract for 10 aircraft. For additional service trials, ABA borrowed the prototype which was converted for freight charter. During the period 11 December, 1948, to 19 March, 1949, ABA made quite a few flights between Nice or Albenga in Italy and Stockholm or Malmö respectively. One nonstop flight between Albenga and Stockholm was recorded for 9 January, 1949, with a flying time of 6 hr 55 min. The ABA trials included 230 flying hours after which the aircraft was returned to Saab and again fitted with 24 seats for further tests and demonstration flights, including a visit to London and a circuit of Finland.

On 12 November, 1949, the first production aircraft made its first flight. No. 2 followed in May 1950 and after that about one aircraft a month was completed except during the holiday season.

On 13 June, 1950, the production Scandia received its type certificate (ICAO Transport Category A) from the Swedish Board of Civil Aviation following completion of 247 flying hours.

The first aircraft ordered by ABA was delivered on 3 October, 1950, after considerable delay according to the original contract. The second followed on 4 November and on 12 December scheduled route trials were started between Stockholm and Luleå in northern Sweden. Scheduled passenger service began on 11 January, 1951, on the Oslo - Gothenburg - Copenhagen route. Later on, Brussels and Amsterdam were included in ABA's Scandia network. In the meantime SAS (Scandinavian Airlines Systems) had been formed by ABA of Sweden, DDL of Denmark and DNL of Norway. As a result the Scandia fleet was based at Oslo's Fornebu airport where the technical service was performed, and the Scandias were finished in SAS livery.

Following delivery of new long-range equipment, notably the Douglas DC-6, the existing long-range DC 4 fleet was put into service on some European routes. Consequently, the SAS Scandia fleet was now mostly used on shorter routes, mainly within Scandinavia. Exceptions were the operations between Stockholm, Riga and Moscow and Copenhagen, Riga and Moscow which were re-opened in May 1956 after 15 years suspension. The Scandia served SAS well during more than six years of scheduled service.

The first export sales were secured in March 1950 when two Brazilian airlines, VASP and Aerovias do Brasil, ordered a total of six aircraft, including the prototype (for VASP). At this time the deliveries to ABA were considerably delayed and as a result the Swedish airline cancelled four aircraft. Ironically, this proved beneficial to both parties; ABA no longer needed as many as 10 aircraft - at least not at the moment - and therefore deliveries to Brazil could start as early as December 1950. VASP, incidentally, already controlled a considerable part of Aerovias and on 21 December, 1950, a complete take over occurred. From late 1951 all Brazilian Scandias operated in VASP colours. Incidentally, in 1954 the prototype was converted into a luxurious executive aircraft for the Brazilian industrialist Olavo Fontoura.

In Sweden the whole Scandia programme had meanwhile reached an extremely critical stage owing to an urgent demand from the Air Force for more production capacity. In 1948 the Swedish Parliament had decided on a 50 percent expansion of the fighter force, 10 extra fighter squadrons. In 1948 a new Saab-designed fighter, the J 29, had made its first flight and the Air Force was planning a series of not less than 600 aircraft. In addition, the Air Force had made a financial contribution to some new production facilities at Saab which in 1950 were partly occupied with the Scandias. Further complicating the matter was the shortage of trained labour and housing.

During the autumn of 1950, the Air Board more or less ordered Saab to stop Scandia production. 'Do what you please with the Scandia but fulfil the J 29 contract', was briefly the stand taken by the Air Force. In this very difficult situation, Saab had little choice but to accept the Air Force conditions despite the heavy investment made in trying to establish Saab as a viable supplier to the airline industry.

Eventually, the Air Board decided to compensate Saab financially through a clever arrangement whereby compensation would be paid in the form of a bonus for each new fighter delivered on schedule.

To rescue Scandia production, Saab first discussed a collaboration with FIAT, which was also anxious to re-establish itself as a producer of transport aeroplanes. These discussions were fairly advanced but eventually fell through. Instead, an agreement was reached on 2 May, 1952, with Fokker of The Netherlands regarding completion of a batch of six aircraft which in 1954 were sold to SAS (four) and VASP (two). Fokker did not have capacity to keep the Scandia line open much longer because the company's resources were fully occupied by its new Fokker F.27 Friendship. And Saab did not consider it profitable to re-open an assembly line in Sweden for the Scandia which could no longer compete with the new generation of pressurized airliners. Fokker completed its first Scandia in April 1954 and the last in October the same year.

In 1957 VASP took over the whole of SAS's Scandia fleet, the last one being

The prototype Scandia at Bromma Airport, Stockholm, in March 1948.
(Saab/Å. Lärkert)

The Scandia was well known for its excellent low-speed characteristics. Seen here is SAS's LN-KLK *Nial Viking.*

delivered in February 1958. VASP also undertook to modify its aircraft to carry 36 passengers in a special four-abreast arrangement. The Scandia was very much appreciated by VASP and its passengers for its reliability. In 1952 the average passenger load factor on its Scandia routes was as high as 93 percent. In 1957 a Scandia became the first airliner to land in the country's new capital Brasilia. Despite acquisition of the turbine-powered Vickers Viscount in 1958, VASP continued to operate the Scandia on a major scale until 1965. The last revenue flight was in fact made on 22 July, 1969. The last aircraft, s/n 90115 (PP-SQR), had served 13 years and 10 months and accumulated 15,683 hr flying time. Altogether only 18 Scandias were built, a small number for a technically very successful aircraft.

In view of its technical success, the question might be asked why the Scandia was not a commercial success. It is not easy to answer this question but its lack of success may have been

Tord Lidmalm *(left)*, **Saab project manager, with Bror Bjurströmer, chief designer, and a model of the Scandia.**

The prototype Scandia with two production aircraft in May 1950. Note the tail strut which was retractable. *(Saab)*

because as an unpressurized airliner it came on the scene too late to compete with highly competitive pressurized types. A pressurized version, the Saab 90B-3, was offered to the airlines in 1949-50, but this was probably too late and it was never built. The marketing efforts in the United States were, incidentally, centred on the B-3 version and discussions were even held with the Glenn L. Martin Company regarding possible licence manufacture. The Martin company was having difficulties at the time with both its 2-0-2 and 3-0-3 models but eventually solved the problems in the modified 2-0-2A and 4-0-4 models. There was no longer any need for a Scandia licence.

Much can and has been said about the engine selected for the production version. It has been claimed that the engine choice was to blame for the fact that the Scandia never became the success it really deserved to be. There may be truth in this but the fact is that when the engine was finally ordered in early 1947 the R-2180 had been specified for several aircraft. In addition to its use in the Scandia, exactly the same version of the engine was selected to power a new Douglas project of similar size designated DC-9 and another version for the Piasecki XH-16 heavy helicopter for the US Air Force. In 1948 Douglas shelved the original DC-9 project and ultimately only Saab used the engine. Altogether, only sixty-five R-2180s were made for Saab and 10 for helicopter use. The first engines were shipped from Pratt & Whitney to Saab in January 1949 and the last in March 1951. It is also indisputable that ABA greatly influenced the choice by specifying the R-2180 for the projected pressurized version. At this time ABA and its intercontinental sister SILA intended to order an initial four Boeing Stratocruisers each powered by four Pratt & Whitney R-4360 Wasp Major engines (they were actually sold to BOAC before delivery). This engine was, in principle, a double R-2180, and as many of the parts were identical, which would have greatly facilitated the spare parts supply. Pratt & Whitney delivered the last spares to VASP in the early 1960s.

90A-2 Scandia

Span 28.0 m (91 ft 10 in); length 21.3 m (69 ft 11 in); height 7.1 m (23 ft 4 in); wing area 85.7 sq m (922 sq ft). Empty weight 9,960 kg (21, 958 lb); payload at 400 km (250 miles) range 3,090 kg (6,812 lb); at 1,300 km (810 miles), 2,035 kg (4,486 lb); maximum take-off weight 16,000 kg (35,274 lb). Maximum speed 455 km/h (283 mph); crusing speed 400 km/h (249 mph); landing speed 130 km/h (81 mph); initial rate of climb 6.5 m/sec (1,300 ft/min); ceiling 8,700 m (28,500 ft); range 2,650 km (1,647 miles).

Saab 90 Scandia production serials

90A-1: (prototype: 90001)
90A-2: 90101–90117 (90001, 90101–90111 were built in Linköping, 90112–90117 completed by *Fokker).*

Individual aircraft

90001	SE-BCA	Prototype. To VASP as PP-XEA, re-registered PP-SQB. To executive aircraft PT-ARS
90101	SE-BSA	Aerovias Brasil PP-XEK, VASP PP-SQF
90102	SE-BSB	Aerovias Brasil PP-XEB, VASP PP-SQC
90103	PP-XEI	Aerovias Brasil, VASP PP-SQE
90104	PP-XEJ	Aerovias Brasil, PP-SQD
90105	SS-BSB	ABA/SAS *Gardar Viking*, VASP PP-SQW
90106	SS-BSD	ABA/SAS *Grim Viking*, VASP PP-SQV
90107	SE-BSH	ABA/SAS *Torulf Viking*, VASP PP-SRB
90108	SE-BSE	ABA/SAS *Jarl Viking*, VASP PP-SRA
99109	SE-BSF	ABA/SAS *Nial Viking*, DNL/SAS LN-KLK, VASP PP-SQY
90110	SE-BSG	ABA/SAS *Sigurd Viking*, DNL/SAS LN-KLL, ABA/SAS SE-CFX, VASP PP-SQY
90111	PP-SQN	VASP
90112	SE-SQQ	VASP PP-SQQ
90113	SE-SQS	VASP PP-SQS
90114	SE-SQT	VASP PP-SQT
90115	SE-SQR	VASP PP-SQR
90116	SE-BSK	ABA/VAS *Arne Viking*, VASP PP-SQZ
90117	SE-BSL	ABA/SAS *Folke Viking*, VASP PP-SQU

One of the 24 Saab 91Ds delivered to the Austrian Air Force in 1964-65. *(Saab)*

Saab 91 Safir (Sapphire)

The Saab 91A Safir was also tested on floats. *(R Wall)*

During 1944 Saab decided to start development of three civil products, an airliner (Saab 90 Scandia), a trainer/tourer (Saab 91 Safir) and a motorcar (Saab 92). The Saab 91 was initially planned as a two-seat trainer but at a later stage a third seat was incorporated. A. J. Andersson was appointed as project manager. In 1939 he had returned to Sweden from Germany where he had been chief designer for the whole range of Bücker trainers from the Jungmann to the Bestmann. The Bestmann ancestry is clearly noticeable in the Safir, although the Safir was designed to be all-metal (the only exception being the aft part of the wing and the control surfaces which were fabric-covered) with a retractable undercarriage. Prototype development started in winter 1944-45 but the work was delayed by a long metal-union strike and the first flight could only take place on 20 November, 1945. The powerplant was a de Havilland Gipsy Major X of 147 hp but despite the limited power, the top speed was as high as 265 km/h (165 mph) which indicates excellent aerodynamic design. The undercarriage was retracted by means of a very simple mechanical spring-loaded arrangement.

Quantity production of the Saab 91A, as the first model was known, began in spring 1946, but during the next two years production was on a limited scale. The market for touring aircraft in this class never really developed as initially expected after the Second World War. Of the Gipsy Major-powered version, 48 examples were produced most of which were eventu-

An all metal beauty, the Saab 91 Safir trainer/tourer made its first flight in November 1945, powered by a de Havilland Gipsy Major X of 147 hp *(R Wall)*

ally sold to the Ethiopian Air Force as primary trainers and to the Swedish Air Force as liaison aircraft.

In May 1947, the Swedish pilot, Count Carl Gustaf von Rosen, made a remarkable long-distance flight from Sweden to Ethiopia in a Saab 91A destined for the Imperial Ethiopian Air Force. Von Rosen, incidentally, served as Chief of that Air Force and that Safir was the 6th aircraft ordered by the Service. Initially, he had planned to make intermediate landings in Rome and Cairo but after some calculations von Rosen concluded that a nonstop flight was possible. The distance of 6,220 km (3,866 miles) would exceed the world record for this class of aircraft. There was no special preparation of the aircraft (the engine was the standard Gipsy Major X driving a two-bladed wooden propeller), except for an extra fuel tank in the rear seat giving a total volume of 947 litres (208 Imp gal) compared to the normal 118 litres. The take-off weight was 1,500 kg (3,370 lb) against the normal maximum take-off weight of 1,075 kg (2,415 lb). Flight tests at Saab with 1,425 kg (3,200 lb) take-off weight produced no problem except extending the take-off run to 900 m (2,950 ft), five times the normal.

A Saab 91A over the English countryside. *(Charles E. Brown)*

In 1949 a more powerful Safir, the Saab 91B, was introduced powered by a 190 hp Lycoming O-435A flat six engine. *(Saab)*

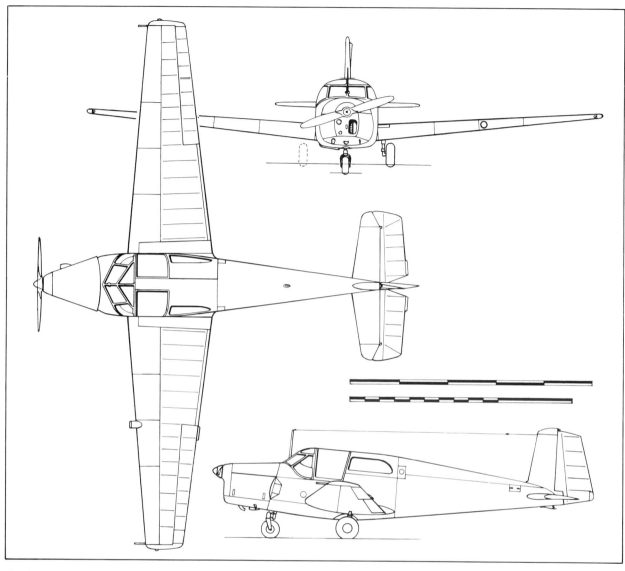

Saab 91A Safir

Apart from encountering sandstorms over the Libyan desert and rainstorms in the Ethiopian mountains, no major problems were met en route. After 30 hr 52 min flying, Count von Rosen landed on the rainsoaked airfield of Addis Ababa; quite an achievement by both pilot and aircraft!

In 1949, Saab announced a more powerful Safir version, the Saab 91B, equipped with a Lycoming O-435A flat six engine of 190 hp. This improved performance significantly.

In 1951 the Swedish Air Force decided to order a new primary trainer and after evaluation of several types, the Saab 91B was selected and 74

A. J. Andersson, project manager for the Safir (LEFT) and Arthur Bråsjö, Lansen project manager.

ordered. Due to capacity problems at Saab, the company decided to subcontract the production to The Netherlands where De Schelde at Dordrecht (which later became part of Fokker) produced a total of 120 Safir aircraft during 1951-55. In addition to the Swedish Air Force, SABENA Belgian Airlines ordered the Safir 91B as did Lufthansa, Air France and the Imperial Ethiopian Air Force. De Schelde also produced a four-seat version, the Saab 91C, in limited numbers.

In 1956 Saab resumed Safir production at Linköping having secured an order from the Royal Norwegian Air Force for twenty-five 91B-2s. Soon thereafter, the Finnish Air Force ordered a batch of twenty of a re-engined version, the 91D powered by a four-cylinder Lycoming O-360-A1A of 180 hp. These were delivered in 1958-59. The same version was also ordered by The Netherlands Government's Airline Pilot School (RLS) which eventually took delivery of as many as nineteen.

In 1960-62 Saab delivered fourteen

In this picture of the Saab 91B cockpit, the dual-control column has been removed. (Saab/Å Anderson)

In 1951 the Swedish Air Force ordered 74 Safir 91Bs (Sk 50). Because of capacity problems at Linköping, the manufacture was subcontracted to De Schelde in The Netherlands. (De Schelde)

A four-seat version of the Safir, the Saab 91C, was also built. *(Saab)*

Saab 91Cs to the Swedish Air Force and ten to Ethiopia followed by a further fifteen to Finland and fourteen for Tunisia. In 1964-65 the Austrian Air Force took delivery of twenty-four Saab 91Ds. In 1966 the final four Safir aircraft produced were delivered to Ethiopia. With a total of 48 Safir aircraft of different versions, Ethiopia became the second largest customer after Sweden for this successful aircraft, 323 of which were produced.

Apart from pilot training, the Safir was used for aerodynamic research work in Sweden (Saab 201 and 202) and in Japan (X1G1-3). An unusual mission undertaken with the original Safir 91A was the successful participa-

Saab 91C Safir

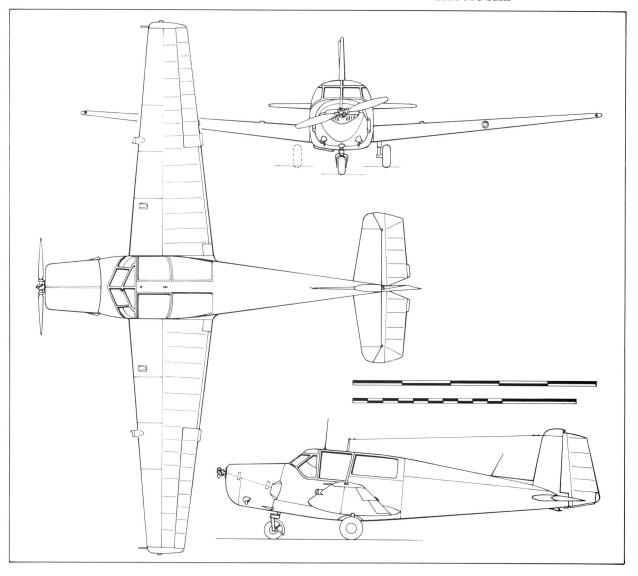

Finland was also a major customer for the Safir, ordering a total of thirty-five. *(Saab)*

tion in late 1951 in the Anglo-Norwegian-Swedish Antarctic Expedition, for which purpose the aircraft operated on floats and skis.

Saab 91A

Span 10.60 m (34 ft 9 in); length 7.80 m (25 ft 7 in); height 2.20 m (7 ft 3 in); wing area 13.60 sq m (147 sq ft). Empty weight 580 kg (1,280 lb); loaded weight 1,075 kg (2,415 lb). Maximum speed 265 km/h (165 mph); cruising speed 248 km/h (154 mph); landing speed 85 km/h (53 mph); initial rate of climb 3.9 m/sec (767 ft/min); ceiling 4,600 m (15,100 ft); range 960 km (597 miles).

Saab 91B

Span 10.60 m (34 ft 9 in); length 7.90 m (25 ft 11 in); height 2.20 m (7 ft 3 in); wing area 13.60 sq m (147 sq ft). Empty weight 730 kg (1,610 lb); loaded weight 1,218 kg (2,736 lb). Maximum speed 275 km/h (170 mph); cruising speed 245 km/h (152 mph); landing speed 90 km/h (56 mph); initial rate of climb 5.35 m/sec (1,052 ft/min); ceiling 6,200 m (20,340 ft); range 1,050 km (653 miles).

Saab 91C

Span 10.60 m (34 ft 9 in); length 7.90 m (25ft 11 in); height 2.20 m (7 ft 3 in); wing area 13.60 sq m (147 sq ft). Empty weight 740 kg (1,630 lb); loaded weight 1,215 kg (2,678 lb). Maximum speed 270 km/h (168 mph); cruising speed 245 km/h (152 mph); landing speed 90 km/h (56 mph); initial rate of climb 5.35 m/sec (1,052 ft/min); ceiling 6,200 m (20,340 ft); range 960 km (597 miles).

Saab 91D

Span 10.60 m (34 ft 9 in); length 8.03 m (26 ft 4 in); height 2.20 m (7ft 3 in); wing area 13.60 sq m (147 sq ft). Maximum speed 270 km/h (168 mph); cruising speed 240 km/h (149 mph); landing speed 90 km/h (56 mph); initial rate of climb 5.0 m/sec (984 ft/min) ceiling 6,100 m (20,000 ft); range 1.125 km (700 miles).

Saab 91 Safir production serials

Prototype
91001 SE-APN To Saab 201 and 202

Saab 91A Safir
91101 SE-AUN, YE-AAG
91102 SE-AUP, IEAF-115
91103 SE-AUR
91104 SE-AYC To Flygvapnet Museum as Tp 91
91105 SE-AZH, ET-
91106 SE-AZI, PP-DIU, SE-AZI
91107 IEAF-101
91108 IEAF-102
91109 IEAF-103
91110 IEAF-104
91111 SAF-91111 (F8-1)
91112 SAF-91112 (F8-2), SE-CDS, D-EMUV
91113 SAF-91113 (F8-3)
91114 SAF-91114 (F8-4), SE-CDT, D-EGIW
91115 SE-AZK, LV-RIG
91116 SE-AZM, IEAF-106
91117 SAF-91117 (F8-)
91118 SAF-91118 (F8-), SE-BNX, to SAF
91119 SAE-91119 (F9-19), SE-BTY, to SAF
91120 SE-AZN, IEAF-105
91121 SAF-91121 (F9-21)
91122 SAF-91123 (F10-22, F8-22)
91123 SAF-91123 (F4- , F8)
91124 SE-BFT
91125 SE-BFU, PH-UEA
91126 SE-BNH, VT-CYU, G-ALCS ntu
91127 SE-BFW, IEAF-116
91128 SE-BNL, SE-BNZ
91129 SE-BNM ABA, IEAF-114
91130 SE-BNN, PH-UEB, OO-HUG, OO-MUG
91131 SE-BNO
91132 SE-BNP, ZP-
91133 SE-AWA, VT-CTS
91134 IEAF-107
91135 SE-AWC, VT-CTT
91136 PH-UEC, G-ARFX
91137 PH-UED, PH-NEP ntu
91138 IEAF-112
91139 PH-UEE
91140 SE-BNU, PH-UEF, OO-JEN
91141 IEAF-113
91142 IEAF-108
91143 PH-UEG
91144 PH-UEH, OO-ANN
91145 IEAF-109, ET-AAN
91146 IEAF-110
91147 IEAF-111
91148 SE-BNT, VT-CZS

Saab 91B Safir (91201–91320 by De Schelde)
91201 SAF-50001, SE-BWB, JA-3055, JAF TX-7101
91202 SAF-50002, (F5-2, F18-81)
91203 SAF-50003 (F5-3, F5-03, F6-71)
91204 SAF-50004 (F5-4)
91205 SAF-50005 (F5-5)
91206 SAF-50006 (F5-6, F7-83)
91207 SAF-50007 (F5-8, F16-95), SE-IGR
91208 SAF-50008 (F5-8, F16-95), SE-IGR
91209 SAF-50009 (F5-9, F18-87, F7-84), SE-IGP
91210 SAF-50010 (F5-10, F12-73), LN-HHS
91211 SAF-50011 (F5-11, F12-71), SE-IIL
91212 SAF-50012 (F5-12, F4-71)
91213 SAF-50013 (F5-13)
91214 SAF-50014 (F5-14)
91215 SAF-50015 (F5-15, F14-92)
91216 SAF-50016 (F5-16, F7-81)
91217 SAF-50017 (F5-17)
91218 SAF-50018 (F5-18, F18-82, F18-83)
91219 SAF-50027 (F5-49, F6-72)
91220 OO-SOK Sabena
91221 SAF-50019 (F5-19, F7-84, F11-72)
91222 SAF-50020 (F5-20, F1-71)
91223 SAF-50021 (F5-21)
91224 SAF-50022 (F5-22)
91225 SAF-50023 (F5-23, F1-72), SE-IGI
91226 SAF-50024 not delivered
91227 SAF-50025 (F5-25, F18-80, F1-80), SE-IGK
91228 SAF-50026 (F5-26, F13-71), SE-IGO
91229 SAF-50027 (F5-27, F11-74)
91230 SAF-50028 (F5-28)
91231 SAF-50029 (F5-29, F16-29, F16-96, F4-73)
91232 SAF-50030 (F5-30)
91233 SAF-50031 (F5-31)
91234 SAF-50032 (F5-32)
91235 SAF-50033 (F5-33)
91236 SAF-50034 (F5-34)
91237 SAF-50035 (F5-35, F15-71)
91238 SAF-50036 (F5-36, F3-74, F7-85, F6-85)
91239 OO-SOL
91240 OO-SOM, IEAF-126
91241 OO-SON, IEAF-127
91242 OO-SOP, IEAF-128
91243 OO-SOQ, IEAF-129
91244 OO-SOR, IEAF-130
91245 OO-SOV, IEAF-131
91246 SE-BYN, OH-SFA, FAF SF-36
91247 SAF-50048 (F5-48, F10-73, F1-)
91248 SAF-50050 (F5-50, F15-74, F16-72)
91249 SAF-50037 (F5-37, F21-77)
91250 SAF-50038 (F5-38, F7-38, F5-38, F1-73, F3-71, F4-75)
91251 SAF-50039 (F5-39, F1-70, F6-73)
91252 SAF-50040 (F5-40, F2-40, F9-40, F11-72, F7-84), NAF 040, LN-HHW
91253 SAF-50041 (F5-41, F1-74)
91254 SAF-50042 (F5-42, F21-72)
91255 SAF-50043 (F5-43, F6-74)
91256 SAF-50044 (F5-44, F12-44, F12-74), SE-IGM
91257 SAF-50045 (F5-45, F15-72, F1-73)
91258 SAF-50046 (F5-46, F17-76)
91259 SAF-50047 (F5-47, F13-72, F1-)
91260 SAF-50051 (F5-51, F13-72)
91261 SAF-50052 (F5-52, F4- F21-74)
91262 SAF-50053 (F5-53, F13-73), SE-IGL
91263 SAF-50054 (F5-54, F17-74)
91264 SAF-50055 (F5-54, F17-74)
91265 SAF-50056 (F5-56)
91266 SAF-50057 (F5-58, F12-72)
91267 SAF-50058 (F5-58, F12-72), NAF 058
91268 SAF-50059 (F5-59, F1-73, F12-72, F21-75)
91269 SAF-50060 (F5-60, F10-72)
91270 SAF-50061 (F5-61, F3-72)
91271 SAF-50062 (F5-62, F1-74, F17-72)
91272 SAF-50063 (F5-63, F17-73, F13-75)
91273 SAF-50064 (F5-64, F4-74)
91274 SAF-50065 (F5-64, F4-74)
91275 SAF-50066 (F5-65, F10-74)

The Saab 91B engine installation was easily accessible. *(Saab)*

Saab 91C Safir (prototype)
91276 SE-BYZ, VH-BQK, VH-AHA, VH-RHG, VH-BHG

Saab 91B Safir
91277 SAF-50067 (F5-67, F16-98)
91278 SAF-50068 (F5-68, F11-73)
91279 SAF-50069 (F5-69)
91280 SAF-50070 (F5-70, F4-70)
91281 SAF-50071 (F5-71, F14-91), NAF 0071, LN-LMY
91282 SAF-50072 (F5-72)
91283 SAF-50073 (F5-70, F4-70)
91284 SAF-50074 (F5-74, F3-71, F1-73), NAF 0074, LN-BII
91285 SAF-50075 (F5-75, F10-71, F21-75)
91286 SAF-50076 (F5-76, F4-73, F16-96), NAF 076, LN-SAL

91287 IEAF-117
91288 IEAF-118
91289 IEAF-119
91290 D-EBAB Lufthansa
91291 D-EBED Lufthansa
91292 (PK-AAK) PK-ASA
91293 PK-AAL
91294 PK-AAM
91295 SE-XAE, F-BHAG
91296 SE-XAF, F-BHAH
91297 SE-XAG, F-BHAI
91298 SE-XAH, F-BHAJ
91299 SE-XAI, F-BHAK
91300 SE-XAK IEAF 120
91301 SE-XAL, IEAF-121
91302 SE-XAM, IEAF-122
91303 SE-XAO, IEAF-123
91304 SE-XAO, IEAF-124

91305 SE-XAP, IEAF-125
91306 SE-XAR, SE-CAE
91307 SE-XAS, SE-CFY

Saab 91B-D Safir (built by De Schelde)
91308 SE-XAT, SE-CFZ, G-AVGS, SE-CFZ (prototype)
91309 SE-XAU, D-EBUC, PH-RJB
91310 SE-XAW, TAF Y 31001

Saab 91C Safir (built by De Schelde)
91311 (SE-CAH), G-ANOK
91312 SE-XAX, SE-CAC, LV-
91313 SE-XAZ, SE-CAD, D-EMUK, PH-BEP
91314 SE-XBA, SE-CAF, OE-DBN, D-EAIW
91315 SE-XBB, SE-CBH, OH-ABC
91316 SE-XBC, (SE-CAG), PK-AAU, PK-ASB
91317 SE-XBD, SE-CBG, PK-AAV
91318 SE-XBE, (EI-AGY), OE-DSA, SE-EDD
91319 SE-XBF, (SE-CBI), I-LUXI
91320 SE-XBG, (SE-CBK), PK-AAW, PK-ASC

Saab 91B-2 Safir
91321 NAF- UA-B 321, G-BKPY T. Newark Air Museum
91322 NAF- UA-D 323
91323 NAF- UA-D 323
91324 NAF UA-E 324
91325 NA UA-F
91326 NAF UA-G
91327 NAF UA-H
91328 NAF UA-I 328, LN-BDI
91329 NAF UA-J 329
91330 NA UA-K 330, LN-SAF

An Air Force student lands a Saab Sk 50. *(Bo Dahlin)*

The Swedish Air Force eventually bought 98 Safirs of different versions including fourteen four-seaters. Many are still in use. *(Bo Dahlin)*

Flygvapnet Safirs are now mostly used for liaison purposes. *(Flygvapnet/F13)*

The Netherlands government's airline pilot training school, Rijksluchtvaartschool (RLS), bought nineteen Saab 91Ds. *(Saab)*

91331 NAF UA-L 331, LN-BEE, SE-101
91332 NAF UA-M
91333 NAF UA-N 333
91334 NAF UA-O 334, SE-CAB
91335 NAF UA-P 335, OO-NOR
91336 NAF UA-Q 336
91337 NAF UA-R 337
91338 NAF UA-S 338
91339 NAF UA-T 339, LN-SAV, SE-IKE
91340 NAF UA-U 340, LN-SAK
91341 NAF UA-V 341, LN-LFK

91342 NAF UA-W 342, LN-SAM
91343 NAF UA-X 343, (LN-HAI), SE-ITF
91344 NAF UA-Y 344, LN-SAO
91345 NAF UA-Z 345, LN-HPD

Saab 91D Safir
91346 SE-CGZ
91347 FAF SF-1
91348 FAF SF-2
91349 FAF SF-3
91350 FAF SF-4, OH-SFF

91351 FAF SF-5
91352 FAF SF-6, OH-SFI
91353 FAF SF-7
91354 FAF SF-8
91355 FAF SF-9
91356 FAF SF-10
91357 FAF SF-11
91358 FAF SF-12
91359 FAF SF-13, SE-IKI
91360 FAF SF-14, OH-SFS, SE-IRY
91361 FAF SF-15

91362 FAF SF-16, OH-SFN
91363 FAF SF-17, SE-IKK
91364 FAF SF-18
91365 FAF SF-19
91366 FAF SF-20 *Tryggve Holm*
91367 PH-RLA
91368 PH-RLB
91369 PH-RLC, SE-IRN
91370 PH-RLD
91371 PH-RLS
91372 PH-RLE
91373 PH-RLF
91374 PH-RLG
91375 PH-RLH
91376 PH-RLK Civil Flying School
91377 PH-RLL
91378 PH-RLM
91379 PH-RLN
91380 PH-RLO
91381 PH-RLP
91382 PH RLR Civil Flying School
91383 PH-RLT
91384 PH-RLU Civil Flying School

Saab 91C Safir
91385 IEAF-132
91386 IEAF-133
91387 IEAF-134
91388 IEAF-135
91389 IEAF-136
91390 IEAF-137
91391 IEAF-138
91392 IEAF-139
91393 IEAF-140
91394 IEAF-141
91395 SAF-50080 (F8-80, F8-51, F18-84, F16 72)
91396 SAF-50081 (F5-81, F4-81, F4-72, F21-71)
91397 SAF-50082 (F5-82, F17-75)
91398 SAF-50083 (FC-71, FC-72)
91399 SAF-50084 (F8-84, F8-52, F18-85)
91400 SAF-50085 (F5-85, F11-75)
91401 SAF-50086 (F5-86, F13-74)
91402 SAF-50087 (F8-87)
91403 SAF-50088 (F5-88, F3-75, F13-79)
91404 SAF-50089 (F5-89, F12-73, F15-73)
91405 SAF-50090 (F5-90, F21-71)
91406 SAF-50091 (F5-91, F21-71)
91407 SAF-50092 (F5-92, F8-53, F18-86)
91408 SAF-50093 (F5-93, F10-75)

Saab 91D Safir
91409 FAF SF-21, OH-SFL
91410 FAF SF-22, OH-SFP
91411 FAF SF-23
91412 FAF SF-24, OH-SFJ
91413 FAF SF-25
91414 FAF-26
91415 FAF SF-27, OH-SFH
91416 FAF SF-28
91417 FAF SF-29, OH-SFC, SE-IKR
91418 FAF SF-30
91419 TAF Y 31002
91420 TAF Y 31003
91421 TAF Y 31004
91422 TAF Y 31005
91423 TAF Y 31006
91424 TAF Y 31007
91425 TAF Y 31008
91426 TAF Y 31009
91427 TAF Y 31010

Tunisia acquired fourteen Saab 91Cs. *(Saab)*

The Imperial Ethiopian Air Force was the second largest customer for the Safir. *(Saab)*

In 1956, the Royal Norwegian Air Force ordered twenty-five Saab 91s. *(Saab)*

91428 TAF Y 31011	91450 AAF 3F-SO	AAF = Austrian Air Force
91429 TAF Y 31012	91451 AAF 3F-SP	FAF = Finnish Air Force
91430 TAF Y 31013	91452 AAF 3F-SQ	IEAF = Imperial Ethiopian Air Force
91431 TAF Y 31014	91453 AAF 3F-SB	JAF = Japanese Air Force
91432 TAF Y 31015	91454 AAF 3F-SC	NAF = Norwegian Air Force
91433 PH-RLV, G-BCFS, LN-MAZ	91455 AAF 3F-SR	SAF = Swedish Air Force
91434 PH-RLW, OO-VOS	91456 AAF 3F-SS	TAF = Tunisian Air Force
91435 PH-RLX, G-BCFT, LN-MAA	91457 AAF 3F-SD	
91436 PH-RLY, G-BCFV, N91SB	91458 AAF 3F-ST	
91437 PH-RLZ, G-BCFW Civil Flying School	91459 AAF 3F-SU	
91438 SE-EDB	91460 AAF 3F-SE	
91439 HB-DBL	91461 AAF 3F-SV	
91440 FAF SF-31, OH-SFB, to Finnish Air Force	91462 AAF 3F-SW	
91441 FAF SF-32, OH-SFK, SE-10C	91463 AAF 3F-SF	
91442 FAF SF-33, OH-SFM	91464 AAF 3F-SX	
91443 FAF SF-34, OH-SFD, to Finnish Air Force	91465 AAF 3F-SG	
91444 FAF SF-35, OH-SEF, to Finnish Air Force	91466 AAF 3F-SH	
	91467 AAF 3F-SI	
	91468 AAF 3F-SJ	
	91469 AAF 3F-SK	
	91470 AAF 3F-SL	

Saab 91C Safir
91445 IEAF-142
91446 IEAF-143

Saab 91D Safir
91447 AAF 3F-SA
91448 AAF 3F-SM
91449 AAF 3F-SN

Saab 91C Safir
91471 IEAF-144
91472 IEAF-145
91473 IEAF-146
91474 IEAF-147

From 1963, the J 29s were equipped with United States Sidewinder (Rb 24) air-to-air missiles. *(Saab/I. Thuresson)*

Saab 29

In the autumn of 1945, the Air Board invited Saab to submit a proposal for a new jet fighter project, tentatively designated JxR. The engine choice was a central issue. The de Havilland Goblin engine already selected for the J 21R was not powerful enough to meet the new performance requirements. In December 1945, the Air Board instructed Saab to base the project on the new de Havilland Ghost (RM 2 in the Swedish Air Force) turbo jet of 5,000 lb (2,270 kp) thrust and 1.35 m (4 ft 5 in) diameter. The basic Air Force demands on the new aircraft were high speed (Mach 0.85-86) and high service ceiling, powerful armament (four 20 mm cannon), excellent manoeuvrability and a rugged design adapted to the special Swedish operating conditions.

Initially a straight-wing configuration was selected but at the end of November 1945 Saab managed, through contacts in Switzerland, to obtain the results of some German aerodynamic research work into swept wings outlining both advantages and disadvantages. This eventually led to a completely new wing configuration. The structural stress and surface finish requirements were other technical challenges. In February 1946, the project, now known as the R 1001, had been 'frozen' in most respects. In addition to extensive wind-tunnel testing, it was decided to undertake practical flight tests with a wing swept at 25 degrees – a suitable compromise meeting the flight safety considerations – fitted to a Saab Safir piston-engined trainer. The inner-wing leading-edge root had more sweep, however, and was thinner. This research vehicle was designated Saab 201 and was subjected to extensive flight testing during 1947. Eventually it became clear that the new fighter was going to achieve at least 1,000 km/h (620 mph).

In the autumn of 1946 most of the remaining basic problems were solved by the Saab engineering team headed by project manager Lars Brising, and three prototypes of the J 29, as the new aircraft was now designated, were ordered. The fuselage was given a barrel-like shape with a central air intake in the nose and a straight air duct to the engine in the rear fuselage. The characteristic shape not surprisingly led to the nickname 'The Flying Barrel'. The rotund fuselage also housed the undercarriage, the armament installation and 1,400 litres (308 Imp gal) of fuel. There was another 700 litres (154 Imp gal) of fuel in the wings plus two 450 litre (99 Imp gal) drop tanks. By sliding the big engine cowling rearwards, the whole engine was easily accessible.

The structure was designed to permit 8 g manoeuvres at maximum speed at low altitude. The first prototype was fitted with extensive instrumentation for automatic recording of surface loads, structural stress and other data. In February 1947, the British test pilot Squadron Leader Robert A. R. (Bob) Moore had been appointed chief test pilot at Saab, the reason being that Sweden had no test pilots with jet experience at this time. And the J 29 *was* very advanced not only for the Swedish aircraft industry. As Bob Moore himself pointed out, this was the first West European jet fighter to have:

a) swept wings, b) all-movable tailplane, c) automatic leading-edge slats, d) full-span ailerons/flaps, e) The D.H. Ghost engine in a single-engine aircraft, and f) a completely new Saab ejection seat.

After the usual taxiing trials had been completed during August, 1 September, 1948, was selected for the first

Lars Brising *(right)*, **project manager for the J 29.** *(Saab)*

Saab J 29F

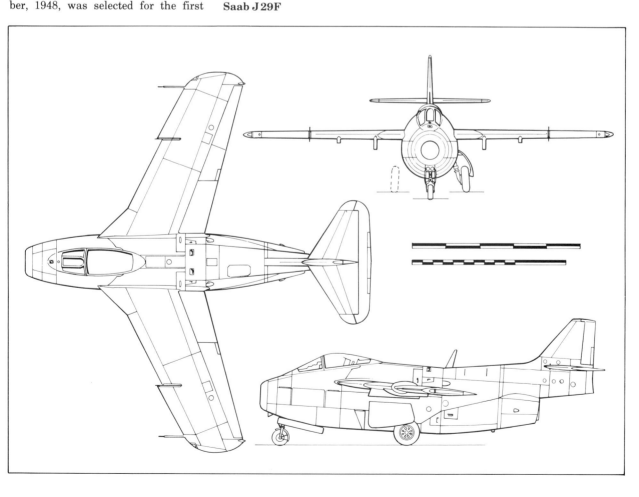

flight. During the flight the speed was so low that the pilot suspected that something was wrong with the air speed indicator but eventually the chase aircraft pilot discovered that, due to a minor mechanical malfunction, the undercarriage doors were open.

Flight testing was not without its problems. On 22 September, 1948, a fied – illustrated the compressed time schedule for the entire programme. The air-brake development was completed in the second prototype in the spring of 1950 without any problems. But there were others; the full-span ailerons in the prototype gave too high a rate of roll, 180 degrees a second. The second prototype was therefore fitted with a

'On the ground an ugly duckling - in the air a swift'. Those were the words of R. A. 'Bob' Moore, the British test pilot after first flying the Saab 29 (J 29) prototype. *(Saab)*

To test the swept wing of the J 29, a Safir was equipped with a scaled down swept wing and designated Saab 201. *(Saab/Wall)*

speed of 890 km/h (553 mph) was achieved at 4,000 m (13,000 ft). At this speed and altitude, however, the use of the wing air-brakes caused unacceptable vibrations, and the brakes were later re-positioned to the fuselage just forward of the undercarriage. The fact that the first 32 production aircraft still had wing air-brakes – somewhat modi- more conventional aileron/flap arrangement which handled beautifully, again according to Bob Moore.

Moore also recalls that in the absence in those days of computers the data from each test flight had to be carefully evaluated before the next step could be taken, a time-consuming operation but necessary in view of the risks from flutter and compressibility. From Mach 0.75 the testing advanced at only Mach 0.01 per flight until approximately Mach 0.8. Here the next major problem occurred in the form of directional snaking which, due to the swept wings, induced Dutch roll which could not be controlled. The design requirement for the J 29 was Mach 0.85 and therefore this problem had to be solved. Following a series of systematic modifications, the test team finally discovered that the trailing angle of the rear fuselage above the jet efflux was wrong, causing breakaway of the airflow on first one side and then the other. The second prototype flew on 29 February, 1949, and the third on 18 August the same year.

The subsequent testing of the prototypes cured the snaking and vibration problems, and there were no doubts that through the J 29 development the Swedish aircraft industry had achieved a very prominent position in fighter aircraft development. Much was also learned about the interference between the aircraft and armament at high speeds which was impossible to calculate and at best uncertain in wind-tunnel testing. The swept wings added to the problems.

In the summer of 1949, the Air Board ordered a fourth aircraft – a production prototype – which flew on 21 July, 1950. It had the tested modifications and was equipped with the more powerful production engine. Performance could be well verified. At 8,150 m (26,740 ft), Mach 0.85 (950 km/h) was achieved without noticeable stability problems or vibrations.

Deliveries began in May 1951, the first Wing to be equipped being F 13 at Norrköping. The introduction into service brought about quite a few serious flight safety problems mainly due to pilots underestimating the difficulties in handling a swept-wing aircraft compared to straight-wing aircraft, particularly on landing. A two-seat trainer version and simulators would have made the conversion training a lot easier. More jet time and improved theoretical tuition before going to the advanced J 29 proved to be the solution.

The production build-up was a major

This unusual view of the J 29 well illustrates the advanced wing shape. *(Saab)*

project and massive investments were made in machinery and tools. The higher precision required more machining than in previous aircraft. For the detail manufacture no fewer than 13,000 tools were produced and 150 major assembly jigs. Of the first version, the J 29A, 224 were produced during 1951-53.

On 11 March, 1953, the first production J 29B was delivered. It carried more fuel in the wings, bringing the total volume from 1,400 litres (308 Imp gal) to 2,100 litres (462 Imp gal). Of this version 361 aircraft were built in two years. In May 1954, a J 29B flown by Capt Anders Westerlund flew a closed-course of 500 km (310 miles) with an average speed of 977 km/h (607 mph), thus beating the previous world speed record held by a United States North

Twenty-four 75 mm air-to-air rockets was another weapon alternative for the J 29. *(Saab/I. Thuresson)*

American F-86 Sabre by 27 km/h. Another world speed record was established in January 1955 when a formation of two S 29C aircraft (the reconnaissance version) flew a 1,000 km (620 miles) closed-circuit with an average speed of 900.6 km/h (559.6 mph). The pilots were Capt (later Major General) Hans Neij and Capt Birger Eriksson.

The S 29C was an unarmed photographic-reconnaissance version, the prototype of which made its first flight on 3 June, 1953. The camera equipment was very extensive for this class of aircraft and up to six vertical and oblique cameras could be carried in the enlarged armament and ammunition bay in the nose of the aircraft. Two alternative camera fits could be carried, one primarily for high-altitude use and one mainly for low-altitudes; and, later on side-looking cameras were installed. The S 29C was much appreciated for its reliability and its endurance of more than 2 hr. The aircraft was also equipped with a Jungner-designed vertical camera sight allowing very high precision on high-altitude missions. A total of seventy-six S 29Cs were built between 1954 and 1956.

On 3 December, 1953, a new version –

The wing of the J 29 was built in one piece and with outstanding surface finish. The J 29A and B versions had automatic slats, the E and F versions a 'sawtooth' wing with fences for improved low-speed handling. *(Saab)*

J 29E – equipped with a modified outer wing, made its first flight. Featuring a 'saw-tooth' wing instead of the leading-edge movable slats, this modification raised the critical Mach number from 0.86 to 0.89. A large number of J 29Bs were modified to J 29E standard and the modified wing was also introduced on the S 29C.

At an early stage, during the flight testing of the No.3 prototype which was allocated to armament testing, the J 29 had already proved to be an excellent weapons platform. It was therefore natural to use the aircraft for air-to-surface missions also. Two attack wings, F 6 and F 7, re-equipped with the J 29B, in its new role re-designated

Special hot-air generators were used to heat the cockpit of the J 29 to a comfortable level while maintaining high readiness in Arctic temperatures down to -40 degrees C (-40 degrees F). *(Saab/Author)*

A pair of J 29Fs over the mountains near the F 4 Wing's base at Östersund. *(Flygvapnet/F 4)*

During September 1954 one J 29 a day was delivered! *(Saab)*

A 29B. This version could carry a variety of ordnance in addition to the four 20 mm cannon with 180 rounds each. The following alternatives could be carried: fourteen 14.5 cm armour-piercing rockets or fourteen 15 cm m/51 rockets. For certain types of missions four 18 cm rockets could be carried. Incendiary bombs could also be carried and these were dropped from very low altitude, 20-25 m. To improve take-off performance JATO bottles were tested but never used in squadron service, the reason being that after-burning was now well underway.

The A 29B era was brief but of great use for the tactical development introduced in the specialized attack aircraft then to come. The A 29B served the Air Force attack units from 1953 until 1957.

A Swedish innovation of great importance to the performance of the J 29 fighter was the development of an afterburner tailor-made to the limited dimensions available in the aircraft. The afterburner was originally conceived by Östen Svantesson of the Air Board and provided a 25 percent thrust increase. The full-scale development took place at Flygmotor which was responsible for production of the RM 2B, as the afterburning version of the Ghost was designated. Flight testing began in March 1954 in an aircraft designated J 29D which also had increased fuel capacity.

A version combining the afterburner engine with the modified wing of the 29E was designated J 29F and flew as early as 20 March, 1954. The incorporation of the afterburner increased the static thrust of the Swedish-manufactured Ghost by as much as 25 percent to 2,800 kp (6,170 lb) which dramatically improved the aircraft's initial rate of climb from approxima-

In March 1954 the first J 29F version, equipped with a Swedish-designed afterburning de Havilland Ghost, made its first flight. Ground running at night was a spectacular sight. *(Saab)*

JATO bottles were tested with heavy loads but were never used on J 29s in service after the availability of the afterburner. *(Saab)*

Seventy-six S 29Cs, the reconnaissance version, were delivered starting in 1953. *(Saab/I. Thuresson)*

An unusual view of eight J 29s of Flygvapnet F 4. *(Flygvapnet/F 4)*

tely 40 m/sec to 60 m/sec (7,875 ft/min to 11,810 ft/min), raising the ceiling from about 13,700 m (44,950 ft) to 15,500 m (50,850 ft), and increasing the top speed from 1,035 to 1,060 km/h (643 to 659 mph). The take-off distance was reduced from 1,350 m (4,430ft) for the J 29B to 790 m (2,590 ft). From 1955 to 1958 a total of 308 J 29B and E versions were modified to J 29F standard, 210 by Saab and the balance by Air Force Central Workshops. 390 RM 2 engines were converted by Flygmotor to RM 2B standard.

Altogether, 661 Saab 29s were delivered from 1951 to 1956. At the end of 1963, all J 29Fs had been modified to carry Sidewinder (Rb 24) air-to-air missiles. The J 29 was the first Saab aircraft to go to war. The background was as follows. The Republic of Congo (now Zaïre) was proclaimed in 1960. The new government had difficulty in establishing an efficient administration and could not maintain law and order. The Province of Katanga broke away and claimed autonomy, supported by Belgian troops. The Congo requested military support from the United Nations. A United Nations force was organized in the Congo in July 1960 and stayed for four years. The UN operations and the air transport of supplies were harassed by Katangan forces and a small fleet of armed light aircraft. One single Fouga Magister equipped with guns, rockets, and bombs caused so much trouble that it was decided that combat aircraft should be added to the UN forces to protect transport aircraft and ground forces. Sweden, Ethiopia and India contributed to the air effort following a UN request in mid-September 1960. The necessary infrastructure to receive the flying unit was provided by the Congo Government, but surveillance radar had to be taken from Sweden.

On 30 September a flight of five J 29Bs, commanded by Col Sven Lampell, took off from Sweden for the 12,000 km (7,460 miles) flight to Leopoldville (now Kinshasa) arriving there on 4 October. On 8 October, the unit, designated F 22 in the Swedish Air Force, redeployed to Luluabourg (now Kananga) in the Province of Kasai. The take-off weight had to be some-

In 1961 Austria ordered an initial batch of fifteen J 29Fs which were delivered in July the same year. In 1963 a further fifteeen were delivered. These were modified before delivery to carry three Vinten cameras instead of two cannon. *(Saab/I. Thuresson)*

Up to seven cameras could be carried by the S 29C. *(Saab)*

what reduced because of the field elevation (700 m above sea level) and high temperatures. The 1,950 m (6,400 ft) runway was just sufficient with full cannon ammunition load and eight 15 cm rockets. The J 29B had been chosen because of lower fuel consumption, important for the long distances in the Congo.

The primary missions were air defence, attacks on air base installations and close support of ground forces. Facilities for reporting and control were practically non-existent but the psychological effect of UN fighters patrolling the area strongly boosted the morale of the Congo troops and chased away Katangan aircraft.

During the Katangan offensive in December 1961, the Swedish aircraft and the Indian Canberras knocked out the Katangan stronghold and the air base at Kolwezi and total air supre-

A trio of J 29s. *(Saab)*

macy was established in a few days. The F 22 unit was deployed to the Kamina base for the remainder of the war, until the autumn of 1963.

Other major attacks were made on Elisabethville (now Lumbumbashi) where vehicle concentrations, trains, staff headquarters and oil dumps were hit.

During 1962 it became evident that reconnaissance aircraft were needed and in November two S 29Cs were allocated. They were, however, airlifted to the Congo by USAF transports. Since by then the Katangan Air Force had been reinforced with combat aircraft and the Ethiopian unit had left the scene, additional Swedish fighters were requested by the United Nations. Four J 29Bs were shipped to the Congo in January 1963 just in time for the final action against Katanga when its air force was wiped out completely. The UN ground operations could now be completed without any air threat. Four J 29Bs were then flown back to Sweden; five were blown up where they stood. The Saab J 29 had completed its UN-mission with top rating and so had the Swedish pilots and the ground crews who kept the aircraft operational. In June 1962 the availability had been 99 percent compared to 54 and 69 for the other air force units participating.

In 1960 Austria decided to acquire combat jets and initially even supersonic fighters were considered. In view of the limited experience of jet combat

Fourteen 15 cm rockets were used against 'hard' targets, four 18 cm rockets against naval targets. As an alternative to drop tanks, Napalm bombs could be carried. *(Flygvapnet Flight Test Centre)*

aircraft, however, an intermediate step was taken. After evaluation of several second-generation fighters including the F-86 Sabre and MiG-17, the Saab J 29F was chosen. An initial batch of 15 was delivered in July 1961. The aircraft were ex-Swedish Air Force machines bought back by Saab and reconditioned before delivery. In the autumn of 1961 another 15 aircraft were acquired. These were modified for combined fighter/attack/reconnaissance missions with two cannon replaceable with the photographic-reconnaisance pack containing three Vinten cameras. The Austrian personnel were trained in Sweden by the Swedish Air Force, the first group of 15 pilots and 40 technicians at the F 15 Wing at Söderhamn. In 1962 a second class of seven Austrian pilots received their conversion training, this time at Säve near Gothenburg. In Austria, the J 29Fs operated from the air bases at Schwechat (Vienna) and Klagenfurt. Later on, the aircraft were also based at Linz and Graz. The principal role of the Austrian J 29s was close support, with patrol of the country's airspace as a secondary role. The aircraft was retired from Austrian military service in 1972 and replaced by another Saab type.

In Sweden the Air Force service life of the J 29 as a combat aircraft ended in May 1967 when the Air Force Wing at Östersund (F 4) flew an '8-plane' farewell formation around the province. But the aircraft remained in Air Force service as a target-tug at Malmslätt near Linköping for another nine years. A special towing winch, MBV-2, was developed for the aircraft, which also carried the Saab BT-23 automatic miss distance indicator and SU-2 flare tracking system fitted to winged and arrow targets. The aircraft also operated in countermeasures training but on 27 August, 1976, completed its last target mission. The last flight took place at a big Air Force 50th anniversary flying display at Malmslätt two days later.

An Austrian Air Force Saab J 29. *(Saab)*

J 29A

Span 11.0 m (36 ft 1 in); length 10.23 m (33 ft 7 in); height 3.75 m (12 ft 4 in); wing area 24 sq m (258 sq ft). Empty weight 4,580 kg (10,097 lb); loaded weight 6,680 kg (14,727 lb). Maximum speed 1,035 km/h (643 mph); cruising speed 800 km/h (497 mph); landing speed 220 km/h (137 mph); initial rate of climb approximately 40 m/sec (7,875 ft/min); ceiling 13,700 m (44,950 ft); range 1,200 km (746 miles).

J 29B/A 29B

Span 11.0 m (36 ft 1 in); length 10.23 m (33 ft 7 in); height 3.75 m (12 ft 4 in); wing area 24 sq m (258 sq ft). Empty weight 4,640 kg (10,229 lb); loaded weight 7,520 kg (16,578 lb). Maximum speed 1,035 km/h (643 mph); cruising speed 800 km/h (497 mph); landing speed 220 km/h (137 mph); ceiling 13,700 m (44,950 ft); range 1,500 km (932 miles).

S 29C

Span 11.0 m (36 ft 1 in); length 10.23 m (33 ft 7 in); height 3.75 m (12 ft 4 in); wing area 24 sq m (258 sq ft). Weights not available. Maximum speed 1,035 km/h (643 mph); cruising speed 800 km/h (497 mph); landing speed 220 km/h (137 mph); initial rate of climb approximately 40 m/sec (7,875 ft/min); ceiling 13,700 m (44,950 ft); range 1,500 km (932 miles).

J 29E

Span 11.0 m (36 ft 1 in); length 10.23 m (33 ft 7 in); height 3.75 m (12 ft 4 in); wing area 24.15 sq m (260 sq ft). Empty weight not available; loaded weight 7,530 kg (16,600 lb). Maximum speed 1,035 km/h (643 mph); cruising speed 800 km/h (497 mph); landing speed 220 km/h (137 mph); ceiling 13,700 m (44,950 ft); range 1,500 km (932 miles).

J 29F

Span 11.0 m (36 ft 1 in); length 10.23 m (33 ft 7 in); height 3.75 m (12 ft 4 in); wing area 24.15 sq m (260 sq ft). Empty weight 4,845 kg (10,680 lb); loaded weight 7,720 kg (17,020 lb). Maximum speed 1,060 km/h (659 mph); cruising speed 800 km/h (497 mph); landing speed 220 km/h (137 mph); initial climb rate approximately 60 m/sec (11,810 ft/min); ceiling 15,500 m (50,850 ft); range 1,100 km (684 miles).

Saab J 29 production serials

J 29 prototypes: 29001, 29002, 29003, 29004 (prototype for J 29E)
J 29A: 21101–29308
J 29B: (later modified to 29E and F) 29325–29683
S 29C 29901–29976

The Saab 32 Lansen was a very popular aircraft and its clean lines are here seen to advantage. *(Saab)*

Saab 32 Lansen (The Lance)

Technical development in the late 1940s and early 1950s not only in the aeronautical field but perhaps even more so in electronics, was nothing short of dramatic. This also greatly affected the form of armed threat and the tactical development to meet it.

During the 1950s a replacement for existing attack, reconnaissance and night-fighter aircraft (Swedish-developed as well as imported) had to be initiated. The first project studies began in late 1946. Twin-jet solutions were in the forefront for several years, a natural trend in view of the fact that both the twin-engined Saab B 18/S 18 family and the de Havilland Mosquito Mk. 19s (J 30 in Sweden) were to be replaced.

After investigating several interesting projects, including a 'flying wing' configuration and an elegant high-wing project known as the 1119 powered by two Flygmotor-built Ghost (RM 2) engines, on 20 December, 1948, the Swedish Air Force decided to go ahead with project 1150, a single-engined transonic two-seater of fairly conventional layout. The 1119 had become too heavy and expensive.

Project manager for the new class of aircraft - several versions were planned - was Arthur Bråsjö. The Saab 32 Lansen was about to be born. As had been the case with the Saab 29, the swept wing (39 degrees) was first tested in reduced scale on a Saab Safir trainer starting in March 1950, the vehicle being designated Saab 202.

Initially Lansen was planned to be powered by a Swedish-developed turbojet, the STAL Dovern II (RM 4) in the 3,300 kp (7,270 lb) thrust class, dry, but an afterburner was being developed. In

The Saab 32 Lansen (A 32) prototype made its first flight on 3 November, 1952, but with the original 'flush' air intakes slightly modified. *(Saab)*

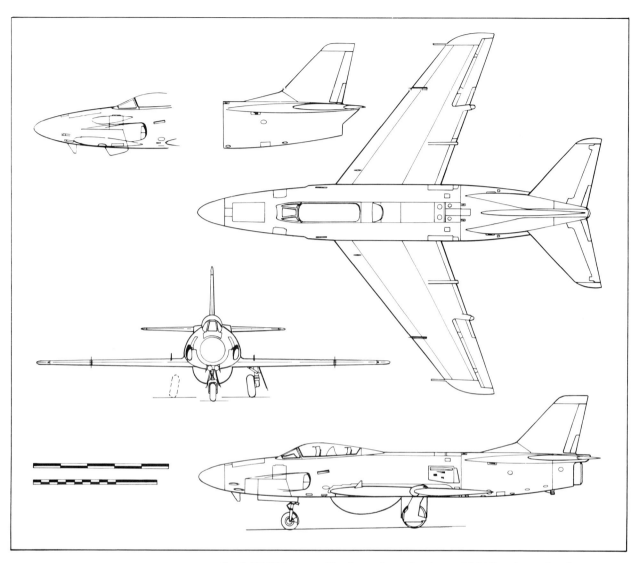

Saab J 32B Lansen. TOP LEFT: front fuselage of S 32C and rear fuselage of A 32A

The swept wing, slat and flap arrangement of Lansen was tested at low speed on a modified Safir (Saab 202). *(Saab)*

November 1952, however, the Air Force decided to abandon Dovern development as well as a later much more powerful engine, the Glan. Instead, a decision was taken to build the Rolls-Royce Avon Series 100 (RM 5 in Sweden) of similar thrust, under licence.

The Avon RA.7R of 3,400 kp (7,490 lb) thrust powered the Lansen prototype from the first flight on 3 November, 1952, with the chief test pilot Bengt R. Olow at the controls. Three additional prototypes were ordered. Lansen was the first two-seat Swedish jet aircraft and the first equipped with a built-in search radar.

The Air Force requirements were very demanding: the aircraft was to be able to attack from a central base any

The A 32 Lansen was the first aircraft to carry the Swedish Rb 04 anti-ship missile, a unique weapon system when it became operational. The Rb 04C version was in quantity production from 1958 until 1964. *(Flygvapnet)*

Lansen was produced in quantity from 1955 until 1960. *(Saab)*

part of Sweden's 2,000 km (1,245 miles) long coast in less than one hour with its armament of four 20 mm cannon, rockets, bombs and missiles, and in all kinds of weather, day and night. Special development effort went into the integration of the electronic and weapons systems – Lansen was Sweden's first true systems aircraft – and also into the aerodynamic configuration. Particular problem areas were the large Fowler flaps and the shape of the rear fuselage, more precisely its integration with the movable tailplane. The final shape was, in fact, determined after studying the flow around the tailplane in a water tank.

The low-speed flight testing in the Saab 202 proved very useful for optimizing the design of the wing. Initially, the first prototype featured flush air intakes but these were modified to a more conventional open configuration before the first flight. On 25 October, 1953, a Lansen prototype exceeded Mach 1.0 in a shallow dive.

The electronic equipment in the aircraft included a PS-43/A search radar, a PN-50A and a PN-51 navigation radar and a PH-11/A radar altimeter. The production aircraft was powered by a Flygmotor-built RM 5A of 3,460 kp (7,630 lb) thrust. With afterburner the thrust increased to 4,700 kp

Flygvapnet had a total of twelve A 32 Lansen squadrons starting in 1956. *(Saab)*

(10,360 lb). The aircraft was also equipped with the Saab BT 9C 'toss' bomb computer and a gyro gunsight was also fitted. The A 32A, as the attack version of Lansen was designated, was armed with four 20 mm cannon in the lower portion of the nose, the gun ports normally closed by electically-operated doors. The external load could comprise either twenty-four 14.5 cm rockets; twelve 18 cm rockets; twelve 100 kg or four 250 kg bombs or for anti-ship missions two Rb 04 missiles. Alternatively two 500 kg incendiary bombs (Napalm) could be carried. A 600 litre (132 Imp Gal) belly tank was normally carried. The integration of the A 32A with the Rb 04, provided the

The varied weaponry that could be carried by the A 32: two Rb 04 missiles; two 500 kg or four 250 kg bombs; twelve 50 kg training bombs; four 20 mm cannon; twelve 18 cm and twenty-four 14.5 cm air-to-surface rockets. Drop tanks or Napalm bombs could also be carried. *(Text & Bilder)*

The J 32B Lansen all-weather fighter was much liked by its pilots. A cleaner-looking aircraft is hard to imagine. *(Saab/I. Thuresson)*

Two A 32s taking off for night flying. *(Flygvapnet/F 6)*

Air Force from 1960 with a unique radar-homing sea-skimming missile which, in a refined version, the O4E, is still a potent weapons system.

Eventually Saab delivered a total of 287 A 32As between December 1955 and June 1957. The aircraft served all five attack Wings (F 6, F 7, F 14, F 15 and F 17).

The second Lansen version to be developed was the S 32C reconnaissance aircraft. It made its first flight on 26 March, 1957. The S 32C carried roughly the same electronic equipment as the A 32A but the built-in cannon armament had been replaced by a PS-431/A search radar with improved performance and a battery of different types of cameras (three SKa 16s, one SKa 15 and two SKa 23s) and a Junger FL S2 optical sight. For night missions, the aircraft could carry twelve m/62 flash bombs. During 1959–60 a total of forty-four S 32Cs were delivered to the F 11 Wing at Nyköping.

During 1955 development began of a night and all-weather fighter version of Lansen. Designated J 32B, it made its first flight on 7 January, 1957. It had 50 percent more engine power and, consequently, true fighter performance. The engine, a Flygmotor-built Rolls-Royce Avon Series 200 (RM 6A in

Sweden), had a dry thrust of 4,790 kg (10,560 lb). With its Flygmotor-designed afterburner, the total thrust reached 6,660 kp (14,680 lb). Due to the larger compressor inlet diameter the air intakes had to be modified as well as the afterburning end. The built-in armament was also much more powerful. The four 30 mm Aden guns used had a considerably higher rate of fire and the muzzle power was 3½ times that of the 20 mm Hispano gun used in the A 32A. For peace-time training purposes, two of the guns were often replaced by 8 mm machine-guns. Externally, the J 32B could carry four Rb 24 (Sidewinder) air-to-air missiles. In addition it could carry two pods each containing nineteen 7.5 cm air-to-air rockets or twelve 15.5 cm rockets intended for ground targets. For special attack missions the aircraft could carry twelve heavy 18 cm rockets.

The J 32B was thus not only a powerful fighter; it also possessed formidable fire power for attack missions. Furthermore, the aircraft was equipped with a new sighting system, the S6A, developed and produced by

Lansen served as a flying testbed for several types of Saab ejector seats. The film strip cut (BOTTOM) shows the first generation rocket-assisted seat and parachute system used in the J 35 Draken. *(Saab/Å. Andersson)*

An unusual view of the Lansen cockpit. Note the extra rear seat windscreen which is necessary in case of canopy emergency release before ejection. *(Flygvapnet/F 17)*

The S 32C reconnaissance version of Lansen was similar to the attack version but its armament was replaced by a modified radar, six cameras, an optical camera sight and flare bombs for night missions. *(Saab/Å. Andersson)*

The crew board a J 32B for a nightfighter mission. *(Saab)*

A Saab A 32 Lansen photographed from another aircraft of the same type. *(Flygvapnet/F 17)*

Saab, which also displayed the target to the pilot in darkness or bad weather. The sighting system was used for firing cannon or rockets against air and ground targets and for missile firing. The S6A worked in conjunction with a new radar sighting system, PS-42A, developed and produced by L. M. Ericsson. In addition, the target could be illuminated by an infra-red camera installed under the starboard wing.

For the first time it became possible to engage targets beyond visual range (BVR) independent of visibility or weather conditions. This capability was probably unique in Europe at that time. The J 32B was equipped with a Saab SA 04 autopilot which made the aircraft a very stable weapons platform in addition to easing the pilot's workload, especially in bad weather. In addition to a modern communications radio

type FR 12 with an F 14 as stand-by, the J 32B had a PN-50/A navigation radar providing course and distance to a radar navigation beacon (PN-51) or radar landing beacon (PN-52). Countermeasures were available but are still classified. The aircraft could also be fitted with dual control for conversion training.

Including the two prototypes, a total of 120 J 32Bs were delivered between

J 32D was the designation of a special version of the J 32B used for target towing. The pod contains a winch. The J 32D, like the J 32E, is still being used for electronic countermeasures training. *(Flygvapnet/F 17)*

July 1958 and May 1960. The first fighter Wing to receive the aircraft was F 12 at Kalmar. F 1 at Västerås followed a year later. Still later, the aircraft also served at the F 4 Wing at Östersund and at F 21 at Luleå, in the north.

In the early 1970s, a total of twenty-four aircraft were transferred to the Target Flying Squadron at Mamslätt. Six aircraft were modified into target tugs as the J 32D equipped with winches for winged and arrow targets, twelve aircraft were allocated for ECCM training under the J 32E designation.

Three J 32Bs also found their way onto the Swedish civil aircraft register in the early 1960s, being used for target operations by Svensk Flygtjänst (later Swedair) based mainly at the missile testing range at Vidsel in Lapland. During seven years, a total of 449 Lansen production aircraft were produced in addition to seven prototypes. It is likely that the J 32B will continue to operate into the next decade.

The STAL Dovern II (RM 4) turbojet originally intended for the Saab 32 Lansen. *(Saab)*

A 32A/S 32C

Span 13.0 m (42 ft 8 in); length 14.9 m (48 ft 11 in); height 4.65 m (15 ft 3 in); wing area 37.4 sq m (403 sq ft). Empty weight 7,500 kg (16,535 lb); loaded weight 13,600 kg (29,983 lb). Maximum speed 1,125 km/h (699 mph); cruising speed 850 km/h (528 mph); landing speed 210 km/h (130 mph); initial rate of climb 60 m/sec (11,810 ft/min); ceiling 15,000 m (49,200 ft); range 1,850 km (1,150 miles).

J 32B

Span 13.0 m (42 ft 8 in); length 14.9 m (48 ft 11 in); height 4.65 m (15 ft 3 in); wing area 37.4 sq m (403 sq ft). Empty weight 8,077 kg (17, 806 lb); loaded weight 11,194 kg (24,678 lb). Maximum speed 1,125 km/h (699 mph); cruising speed 850 km/h (528 mph); landing speed 250 km/h (155 mph); initial rate of climb 100 m/sec (19,685 ft/min); ceiling 16,000 m (52,500 ft); range 2,000 km (1,245 miles).

Saab 32 Lansen serials

Prototypes: 32001, 32002, 32003, 32004; B: 32501, 32502; C: 32901
J 32B: 32501–32620
S 32C: 32901–32945
J 32D: Converted J 32B
J 32E: Converted J 32B

Civil registration

32567 as SE-DCL
32611 as SE-DCM
32616 as SE-DCN

The Saab 210 was a specially designed research aircraft used to test the very advanced 'double-delta' wing layout selected for the Saab 35 Draken supersonic fighter. Three different air intake configurations were tested. *(Saab)*

Saab 35 Draken (The Dragon)

Draken project manager, Erik Bratt, talking to the Saab chief test pilot, Bengt R. Olow, after the first flight. *(Saab)*

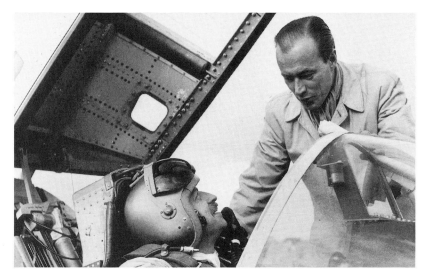

Initial project studies for a replacement of the J 29 'Flying Barrel' had begun in the autumn of 1949, and in October that year Saab received an initial study contract for an aircraft tentatively designated the 1200. Fast, high-flying bombers were considered the greatest threat at that time and a top speed of Mach 1.5 was required. Initially two different projects were analysed, one swept-wing fighter (the 1220) and one delta-winged (the 1250). Much pointed in the direction of the delta configuration but after two accidents in the United Kingdom a certain scepticism had proliferated in Europe, though not so at Saab where Erik Bratt was the project manager. In order to gain some practical experience and above all test the low-speed characteristics, a 70 per cent experimental aircraft, known as the Saab 210, was built. It was ready to fly in December 1951 but owing to extremely bad weather the first flight did not take place until 21 January, 1952. The chief test pilot Bengt R. Olow was very pleased with the landing characteristics, of critical importance to flight safety. Being powered by a small Armstrong Siddeley Adder engine with only 5-10 percent of the new fighter's engine thrust, flight performance was of course limited. (Speed at sea level was 555 km/h/345 mph). But the testing proved extremely useful and a total of 887 flights totalling 286 hours were actually made. Flights were made with three different air intakes and there were even some aerodynamic weapons tests.

The Swedish Air Force accepted the 1250 project for full-scale development at the beginning of 1952 and in April three prototypes were ordered under the Air Force designation J 35. Later the name Draken was added. Draken was not a pure delta-winged aircraft; instead an extraordinary 'double delta' configuration was chosen. This was developed to meet both the high-speed and low-speed requirements and short take-off and landing distances. The inner wing was given as much as 80 degrees sweep while the outer wings at 60 degrees sweep provided for the lower speed range. The inner wing featured low drag but was broad and thick, permitting ample space for air-intakes, main undercarriage, fuel and cannon armament. The inner wing was integral with the fuselage which gave many advantages from a stress point of view. The wing's trailing edge was designed for a combined elevator/aileron function. The air-brakes were positioned around the rear fuselage.

The first prototype flew for the first

The Svenska Flygmotor-produced RM 68B had an afterburning thrust of 6,535 kp (14,395 lb). The afterburner was completely designed and developed in Sweden. *(Volvo Flygmotor)*

OPPOSITE TOP: Sweden's first supersonic fighter, the Saab 35 (J 35) lands after its maiden flight. *(Saab)*

OPPOSITE BOTTOM: Five Draken prototypes in 1956. They were all powered by Swedish-built Rolls-Royce Avon Series 100s (RM 5 in Sweden) but the production version used the more powerful Series 200 (RM 68) engine. *(Saab/I. Thuresson)*

Draken production deliveries began in 1959 and the following summer this impressive J 35A line could be seen at the F 13 Fighter Wing at Norrköping. *(Saab)*

In this picture the J 35F carries two Rb 27s and two Rb 28s (HM-58). The latter is IR-homing. Note the IR seeker under the nose. *(Saab/I. Thuresson)*

One of Draken's two 30 mm cannon being loaded. *(Saab/Å. Andersson)*

time on 25 October, 1955, with Bengt Olow at the controls. The second and third aircraft followed in rapid sequence in January and March 1956 respectively. The flight-test programme was very extensive, with heavily instrumented prototypes and automatic flight-test data evaluation. A new ejector seat was developed by Saab and extensively tested on the ground and in the air.

The three prototypes were all equipped with the same engines as Lansen, the Flygmotor-built Rolls-Royce Avon Series 100 (RM 5A). For the production aircraft, however, the more powerful Series 200 (RM 6B) of 4,890 kp (10,780 lb) dry thrust was specified.

On 26 January, 1956, the prototype exceeded Mach 1.0 in a climb and without afterburner.

In August 1956, the first production contract was signed by the Air Board and on 15 February, 1958, the production prototype – the fifth aircraft to be built – made its first flight powered by an RM 6B engine with a total afterburning thrust of 6,535 kp (14,400 lb).

Production deliveries started late in 1959 and until 1962 a series of ninety J 35As was completed, including prototypes. From the 66th aircraft, a new rear fuselage configuration was introduced. There was a new afterburner which required a new tail cone with a retractable tailwheel to facilitate aerodynamic braking using a nose-high position after the touchdown to save wear of brakes and tyres. During 1959-60, twenty-five J 35As were converted into two-seat Sk 35C trainers for the

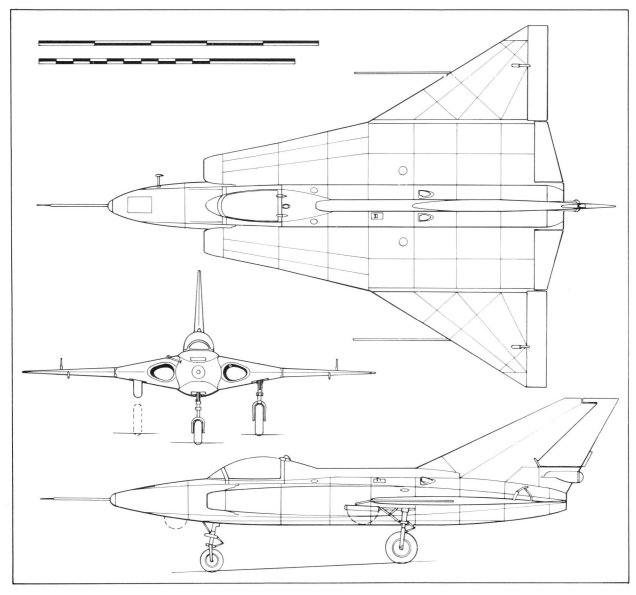

Saab 210

Conversion Training School established at the F 16 Wing at Uppsala. Several flight simulators were also acquired.

The J 35A was equipped with radar produced by L. M. Ericsson (PS-02, PN-793/A) based on the French CSF Cyrano. A Saab 6B sighting system was used. The main armament was four IR-homing Rb 24 (Sidewinder) missiles but the aircraft also carried two 30 mm Aden type cannon plus air-to-surface rockets. The J 35A version was used by the Fighter Wings at Norrköping (F 13) and Uppsala (F 16).

On 29 November, 1959, a new Draken version, the J 35B, made its first flight. It had a greatly developed radar and sighting system, the X-band pulse radar having been developed by L. M. Ericsson and the new collision-course sighting system, S 7, by Saab. The armament was increased to include two pods containing a total of thirty-eight 7.5 cm air-to-air rockets. A J 35B prototype, on 14 January, 1960, exceeded Mach 2.0 in level flight. A total of 73 J 35Bs were delivered in 1962-63 to the fighter Wings F 16 (Uppsala) and F 18 (Tullinge) near Stockholm. The latter unit organized the well-known aerobatic team, Acro Deltas.

A further development of Draken, the J 35D, with an even more powerful engine, made its first flight on 27 December, 1960. Powered by a Flygmotor-built Avon Series 300 (RM 6C) with an afterburning thrust of 7,800 kp (17,200 lb) this version had much improved performance, including an initial rate of climb of approximately 250 m/sec (49,200 ft/min). The J 35D carried more fuel, had a rocket-propelled ejector seat, extended and sharper air intake lips, and other refinements. The avionics included the Saab S 7A sighting system, the LME PS-03 radar and the Saab FH-5 autopilot.

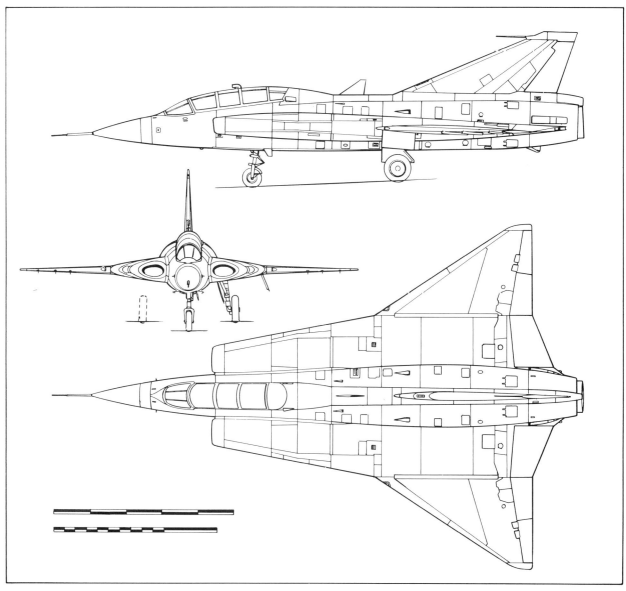

Saab SK 35C Draken

During 1962–63 a total of 120 J 35Ds were delivered, serving with the fighter Wings F 13 (Norrköping), F 3 (Malmslätt), F 4 (Östersund), F 10 (Ängelholm) and F 21 (Luleå).

Draken was also developed in a reconnaissance version, S 35E, the prototype of which flew on 27 June, 1963, for the first time. The limited space in the fuselage nose made necessary a new type of camera which was developed to Swedish specifications in France by OMERA/Segid as the SKa 24, nine of which were carried by the S 35E, five in the nose and four in the wings instead of the guns. The S 35E also pioneered the development of a new infra-red reconnaissance system. The active IR system was developed by the American company EG & G. Edgerton, Graier and Germeshausen, and the cameras by Vinten. The equipment, which was carried in a Swedish pod, was well received and is now standard in newer Swedish reconnaissance aircraft. Of the S 35E a total of sixty were delivered, the first 32 newly built, the remainder converted J 35Ds.

The S 35E had no armament, its escape tactics were to use supersonic speed at low altitude, and this was possible with drop tanks fitted.

Considerable development effort was required for the next Draken version, the J 35F, which was really a new weapons system incorporating the United States Hughes HM 55 and HM 58 Falcon missile system. The missiles were to be built in Sweden under licence by Saab. The aircraft itself was directly derived from the J 35D, with the same engine, but included a host of new features such as:

- A new LME-developed target acquisition radar (PS-01) with a Saab S 7B collision-course sighting system

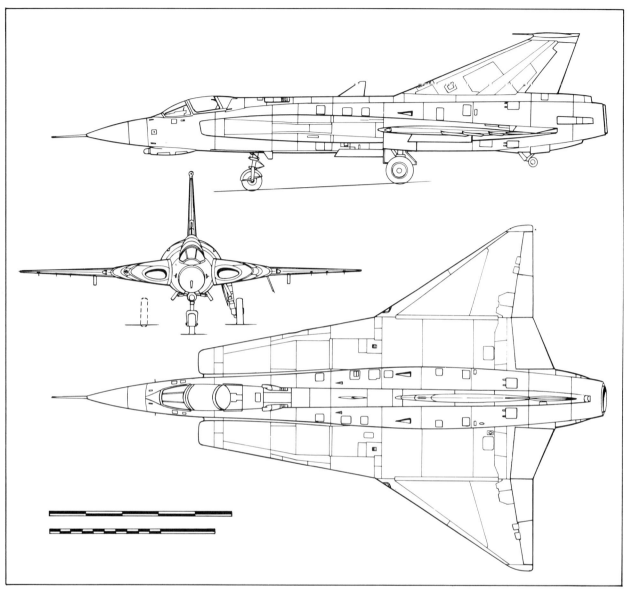

Saab J 35F Draken

- Missile support systems for radar and IR weapons (Rb 27 and Rb 28 in Sweden)
- IR search and track set below the nose radome
- Improved radio communication, navigation and IFF equipment
- Integration with the Air Force's STRIL 60 computerized air defence control and reporting system

The development work on the J 35F began in January 1959. This was a real challenge for the Air Board and the industrial companies involved in this advanced integration of missiles, radar and sights. The work was managed by

The camera-packed nose of the S 35E. *(Saab)*

The Saab J 35F Draken in its wartime environment, operating from a road. *(Saab/H. O. Arpfors)*

a special group headed by Colonel (Engineer) Gunnar Lindqvist (today Major General and head of the Air Material Department in the Defence Material Administration, FMV). In systems performance the J 35F became the most advanced fighter of European design for many years. The most important new capability was missile attack against targets in cloud with the Rb 27 radar missile. These were complemented by the Rb 28 and the Rb 24 IR missiles. The missiles, radar PS-01 (and a later variant PS-11) and the sighting system comprise an integrated computerized unit which takes over the complex tasks imposed by an attack, the pilot's responsibility being greatly eased. One cannon is retained for use against targets taking violent evasive action. The new PS-011 radar is 'reinforced' against enemy ECM.

Compared to the earlier J 32B weapons (Rb 24), the Falcon collision-course weapons have increased the lethal range of the fighter aircraft from some 2 km at low altitude to about 3 km, and at high altitude about 6 km to 10 km.

Heaviest Draken ever: six 1,000 lb bombs and two 1,275 litre (280 gal) drop tanks were required for the Danish attack version ordered in 1968. *(Saab/I. Thuresson)*

Two J 35Fs taking off. The Rb 27 radar-homing missile is the main armament here. *(Flygvapnet/F 13)*

The Saab 35XS, an export version of Draken for Finland derived from the J 35F. Twelve such aircraft were assembled by Valmet. *(Finnish Air Force)*

DK-269 and DK-261 of the Finnish Air Force. These were of the Saab 35XS export version of the J 35F. *(Saab)*

For attack missions the J 35F can carry twelve 13.5 cm Bofors rockets or two pods with a total of thirty-eight 7.5 cm rockets with very high precision. In Draken's early service life the special flight characteristics of the delta-winged aircraft led to a number of accidents caused by 'superstall', the uncontrolled violent nose-up, nose-down attitude inducing dramatic loss of speed and altitude, and special training was introduced to overcome this serious flight safety problem. A number of 35C Draken trainers were equipped with anti-spin chutes which recovered the aircraft from the uncontrolled attitude. The Swedish Air Force has given systematic superstall training since 1971 as a result of which such accidents have now almost been eliminated.

With Draken fighters the Swedish Air Force has further developed the capability of the pilots to use specially prepared stretches of roads as temporary air bases. This capability has dramatically increased the survivability of the Air Force units on the ground.

Of the J 35F version, 230 were produced during the period 1965-72 and the aircraft served with F 13, F 1, F 12, F 16, F 17 and F 10 Wings.

Draken has also been exported. The first export order was secured in 1968 when the Danish Air Force ordered an initial batch of a new attack version know as the F-35 in the Danish Air Force. Later, a special reconnaissance version was ordered, the RF-35, as well as the TF-35 two-seat trainer version. Most of the aircraft were delivered during 1970-72 but more trainers were ordered in 1976 bringing the total Danish Draken procurement to 51. Most of these were still in service in 1988. In the mid-1980s the Danish Air Force undertook a major modification programme of Draken avionics.

The Danish Drakens were modified to NATO standard in weapon pylons and other features and carried extra-large drop tanks (two of 1,275 litres) as well as arrester hooks for emergency landings. Different electronic equipment was also specified, and heavy armament. The maximum take-off weight was increased to 16.5 tonnes (36,376 lb). The fact that the F-35 is still able to get off the ground in about 1,200 m shows the excellent performance of the Draken platform.

In 1970, Finland decided to acquire a batch of twelve 35S (S = Suomi), a slightly modified version of the J 35F, for final assembly by the Finnish aircraft concern, Valmet, at Halli. To facilitate conversion training, Finland leased from the Swedish Air Force six J 35Bs in 1972 and four years later the

The Ericsson PS-01 radar in the J 35F. The PS-11 variant was further 'reinforced' against ECM. *(Ericsson)*

Units of the PS-01 radar plus the antenna and drive. *(Ericsson)*

aircraft were bought together with six J 35Fs and three Sk 35C trainers. The last of the Finnish 35Ss were delivered in 1975 to the Draken fighter Wing at Rovaniemi in Lapland. In March 1984, the Finnish and Swedish Air Forces signed an agreement under which Finland acquired an 'additional number' of Draken aircraft plus ground support equipment. In 1985, Finland exercised an option in the agreement to order further armament equipment for Draken which is now an important part of the Finnish Air Force. Two Wings are currently flying the Swedish aircraft.

A third country, Austria, in May 1985, signed a contract for twenty-four refurbished J 35D fighters. Deliveries started late in 1987 and again Austrian

Low-level reconnaissance over the Swedish Baltic archipelago by an S 35E Draken. *(Saab/I. Thuresson)*

pilots and technicians were trained in Sweden, the pilots by the Swedish Air Force.

In 1985, the Swedish Air Force decided to modify and refurbish approximately sixty J 35Fs into a new version, the J 35J, with more powerful missile armament, longer range and up-dated avionics. These aircraft are expected to serve in the Swedish Air Force well into the mid-1990s, giving a basic service life of 35 years for this remarkable Swedish fighter of which 604 were produced including prototypes and conversions.

J 35A/Sk 35C

Span 9.42 m (30 ft 11 in); length 15.2 m (49 ft 10 in); height 3.89 m (12 ft 9 in); wing area 49.2 sq m (529.6 sq ft). Loaded weight 9,000 kg (19,842 lb). Maximum speed Mach 1.5; cruising speed Mach 0.9; initial rate of climb 200 m/sec (39,370 ft/min); landing speed 215 km/h (133 mph); ceiling 15,000 m (49,200 ft); range 2,750 km (1,709 miles).

J 35B

Span 9.42 m (30 ft 11 in); length 15.34 m (50 ft 4 in); height 3.89 m (12 ft 9 in); wing area 49.2 sq m (529.6 sq ft). Loaded weight 9,000 kg (19,842 lb). Maximum speed Mach 1.8; cruising speed Mach 0.9; initial rate of climb 250 m/sec (49,200 ft/min); ceiling 15,000 m (49,200 ft); range 2,750 km (1,709 miles).

J 35D/S 35E/J 35F

Span 9.42 m (30 ft 11 in); lenght 15.34 m (50 ft 4 in); height 3.89 m (12 ft 9 in); wing area 49.2 sq m (529.6 sq ft). Loaded weight 11,000 kg (24,250 lb). Maximum speed Mach 2+; cruising speed Mach 0.9; initial rate of climb 250 m/sec (49,200 ft/min); landing speed 230 km/h (143 mph); ceiling 20,000m (65,600 ft); range 2750 km (1,709 miles).

Saab 35 Draken production serials

J 35A prototypes:	35001–35004
J 35A:	35005–35090 (from 35066 with a new longer fuselage)
Sk 35C:	35800*–35825
J 35B:	35201–35273
J 35D:	35274–35393
S 35E:	35901–35931, 35932–35960 (latter series modified J 35Ds)
J 35F:	35401–35630
F-35, RF-35 TF-35:	351001–351020, 351101–351120, 351151–351161 (Denmark)
35XS:	serials not available for Finnish versions.

*35800 retained by Saab for research and development

The Saab 105, with two Turboméca Aubisque engines, made its first flight in July 1963. Initially a private-venture project, it was adopted by the Swedish Air Force in December 1961. *(Saab/I. Thuresson)*

Saab 105

A more powerful version of the Saab 105 equipped with two General Electric J-85-17B engines made its first flight in April 1967. Their accessibility is being demonstrated. *(Saab/Å. Andersson)*

In the autumn of 1958 the Swedish Air Force started to look for an 'all-through' jet trainer to replace both the Vampire Trainer and some piston-engined trainers. Saab had meanwhile conducted studies into a combined executive jet and trainer aircraft featuring a delta wing. Soon it transpired that this configuration was not suitable for the Air Force roles, especially as the preliminary specification also called for secondary attack missions.

In mid-1960 the Saab 105 private-venture project was submitted to the Air Force. It was a high-winged, moderately swept, twin-engined monoplane with instructor and pupil seated side by side beneath a large one-piece Perspex canopy. Ejector seats were installed. Two French Turboméca Aubisque by-pass engines each of 742 kp (1,635 lb) static thrust had been chosen to power the new aircraft. Ragnar Härdmark had been appointed project manager.

A large number of Flygvapnet pilots have received their jet training in the Saab 105 (Sk 60) twin-jet, side-by-side trainer. The ejector seats are of Saab's own design. *(Saab/Å. Andersson)*

On 16 December, 1961, the Government authorized Saab to 'continue the development of the Saab 105 twin-jet trainer which should also include attack capability' and in April 1962 a preliminary contract was signed between the Air Board and Saab for procurement of 130 aircraft on the condition that 'the coming flight testing showed that the aircraft could meet the requirements'. A Government decision in March 1964 formally authorized the Air Board to go ahead with the acquisition of the new Sk 60, as the aircraft had been designated.

Manufacture of the two prototypes began in 1962 and on 1 July, 1963, Karl-Erik Fernberg, the test pilot, was ready to go. The early flight testing showed that both the air intakes and the outlets had to be considerably modified, including the wing root beneath which the intakes were situated. The second prototype flew for the first time in June 1964. The third aircraft produced, which was the first production aircraft, made its first flight

Three main versions of the Saab 105 were developed for the Air Force including the Sk 60A trainer, the Sk 60B trainer/attack and Sk 60C attack and reconnaissance aircraft. The 'C' features a characteristic extended nose housing a panoramic camera and an IR seeker. It is seen here being loaded with twelve 13.5 cm air-to-surface rockets. *(Saab/I. Thuresson)*

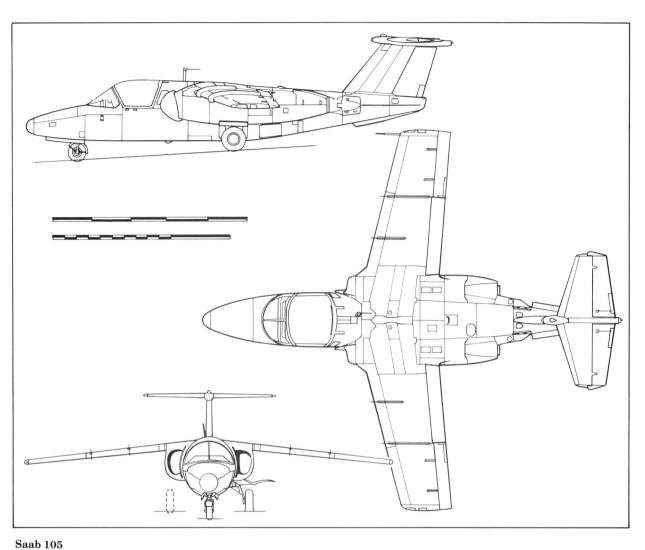

Saab 105

Sk 60Cs taxi-ing to take-off at a wartime air base in northern Sweden. *(Saab/I. Thuresson)*

In 1968 the Austrian Government ordered twenty Saab 105XTs followed by another twenty in 1972. They are mainly used for advanced training and ground support. Reconnaissance pods can also be carried. *(Saab)*

As an alternative to rockets the Sk 60C can carry gun-pods containing 30 mm cannon. They are of the same type as that used by the AJ 37 Viggen. *(Saab/I. Thuresson)*

in August 1965. At this point, the Air Board had increased its order from 130 to 150 aircraft.

In April 1966 the Air Force Flying School (F 5) at Ljungbyhed received its first Sk 60. During the Sk 60's early Service life, the engines (RM 9s) developed some problems which called for extensive modifications. With this programme completed, the reliability became completely satisfactory. On 17 July, 1967, the first students began their training on the Sk 60 and by 1968 all 150 aircraft had been delivered. For the Swedish Air Force several Sk 60 versions were developed from the original Sk 60A trainer version.

Sk 60B is an attack version adapted to carry pylons for two 30 mm Aden cannon or up to twelve 13.5 cm rockets, and some sixty aircraft were modified to B standard from 1970. In addition, about 20 aircraft were modified into Sk 60Cs, a combined photographic-reconnaissance and attack version, the former version having an extended nose with a Fairchild KB-18 panoramic camera and an IR search unit.

A fourth version, the Sk 60D, was modified into a four-seater (the ejector seats being removed) and equipped with commercial nav-com equipment for airline training of reserve officers. Most of the modification programmes were undertaken by the government maintenance facilities (CVM) at Malmslätt. The Sk 60 has continuously served with a total of five Wings and will continue to serve for many years to come as a result of a structural modification programme started in 1988 and involving as many as 135 aircraft. The fact that so many aircraft still serve in the Swedish Air Force after 15–20 years of service as a trainer and low-level strike aircraft of the original 150 produced is proof of the inherent flight safety of this design.

In 1967 a special export version of

Since 1972 the Saab 105OE, as the Austrian version is designated, have been used for patrolling Austrian airspace in the absence of specialized jet fighters. From 1988, however, the 'policing' of the airspace will be taken over by twenty-four Saab 35OEs, refurbished ex-Flygvapnet J 35Ds. *(Saab)*

In 1988 the Swedish Air Force decided to refurbish and modernize as many as 135 Sk 60s to serve into the next century. This example is seen with 30 mm gun-pods and a nose camera. *(Saab/I. Thuresson)*

The Saab 105XT, as a new version was designated, was extensively tested with different weapons including heavy bombs and missiles. Here it launches a salvo of 13.5 cm rockets, a standard alternative in the Swedish Air Force. *(Saab/I. Thuresson)*

the Saab 105 was developed with about 70 percent more power and much higher performance. Designated the 105XT, the prototype made its first flight on 29 April, 1967, equipped with two General Electric J-85-17Bs each of 1,293 kp (2,850 lb) st thrust. Instead of the 700 kg of external stores that can be carried by the Sk 60, the 105XT can carry up to 2,000 kg (4,410 lb). Owing to the higher fuel consumption, the internal fuel volume was increased from 1,400 to 2,050 litres (308 to 450 Imp gal) with provision for two 500 litre (110 Imp gal) drop tanks. The top speed increased from 770 km/h (478 mph) to 970 km/h (603 mph) and to cater for the higher performance the wing structure was strengthened.

In July 1968, the Austrian Government ordered twenty aircraft based on the XT version and designated 105OE. Later, twenty additional aircraft were ordered. All forty aircraft were delivered during 1970-72 and served mainly for training and ground support but, Austria lacking fighter aircraft, they were also used down to the present day for controlling Austrian airspace.

An Sk 60C over the mountains in northern Sweden. *(Saab/I. Thuresson)*

Sk 60A

Span 9.50 m (31 ft 2 in); length 10.50 m (34 ft 5 in); height 2.70 m (8ft 10 in); wing area 16.30 sq m (175 sq ft). Empty weight 2,510 kg (5,534 lb); loaded weight 4,050 kg (8,930 lb). Maximum speed 770 km/h (478 mph); cruising speed 710 km/h (441 mph); landing speed 165 km/h (103 mph); initial rate of climb 20 m/sec (3,940 ft/min); ceiling 13,500 m (44,300 ft); range 1,780 km (1,106 miles).

Sk 60B/C

Span 9.50 m (31 ft 2 in); length (Sk 60C) 11.00 m (36 ft 1 in); height 2.70 m (8 ft 10 in); wing area 16.30 sq m (175 sq ft). Empty weight 2,510 kg (5,534 lb); loaded weight 4,500 kg (9,920 lb). Maximum speed 765 km/h (475 mph); cruising speed 700 km/h (435 mph); initial rate of climb 17.5 m/sec (3,445 ft/min); landing speed 165 km/h (103 mph); ceiling 12,000 m (39,400 ft); range 1,780 km (1,106 miles).

Saab 105XT

Span 9.50 m (31 ft 2 in); length 10.50 m (34 ft 5 in); height 2.70 m (8 ft 10 in); wing area 16.30 sq m (175 sq ft). Empty weight 2,515 kg (5,545 lb); loaded weight (with armament) 6,500 kg (14,330 lb). Maximum speed 970 km/h (603 mph); cruising speed 800 km/h (497 mph); landing speed 165 km/h (103 mph); initial rate of climb 77 m/sec (15,000 ft/min); ceiling 13,700 m (44,950 ft); range 2,760 km (1,715 miles).

Saab 105 production serials

Prototypes: 60001, 60002
Sk 60A, B, C and D: 60002-60150
105OE: 105401-105440

The design mock-up of the Saab 37 Viggen was very detailed. It was shown publicly for the first time on 5 April, 1965. The missile mock-up is the Saab RB 05A radio-command air-to-surface weapon. *(Saab)*

Saab 37 Viggen (The Thunderbolt)

Preliminary studies of designs intended to replace the Swedish Air Force's existing attack, reconnaissance and fighter aircraft began in the early 1950s. After evaluating a large number of more or less specialized attack and fighter projects, including further developments of the Draken family, the idea of a multi-role aircraft using complementary equipment for the attack, reconnaissance and fighter roles in order to keep the development, production and life-cycle costs down, was accepted by the Air Force.

In February 1961, the Air Board submitted its detailed requirements to the industry. During the year, the Saab project team worked hard on Aircraft 1500, as the project was labelled. In June 1961 the Supreme Commander had approved the Air Force Specifications for the new aircraft and in December 1961 a Government decision to develop Aircraft System 37 was taken. The system would include the AJ 37 attack aircraft forming the multi-role 'platform' to be followed by the S 37 reconnaissance version and the JA 37 fighter version.

In December 1962, the Air Board and Saab made a detailed Press presentation of System 37 and for the first time it was disclosed that the new aircraft would have an outstandingly novel aerodynamic configuration. This extremely advanced configuration used a fore-plane fitted with flaps, in combination with a main delta wing. The delta-shaped fore-plane is placed in front of and slightly above the main wing, and thus serves as a lift generator, making possible very low landing

Dramatic view of the 'fighting face' of Viggen. *(Saab/W. Linder)*

Viggen turning briskly immediately after lift off using full afterburner thrust. *(Saab)*

In order to meet the requirements for short landing performance, Viggen is equipped with an effective thrust reverser making possible landings in about 500 metres. *(Saab/Å. Andersson)*

speeds. The unconventional wing configuration enabled the aircraft to meet the Air Force's stringent STOL requirements without resorting to expensive variable geometry or lift-configured engines. Low sensitivity to turbulent air was another great advantage.

To enable the aircraft to land on very short runways, Saab developed a thrust-reverser integrated with the rear fuselage, still the only one of its kind in a single-engined aircraft. The thrust reverser is efficient even on icy runways. By using autothrottle on the approach, a head-up display (HUD) to give touchdown precision in combination with an undercarriage dimensioned for carrier-type landings, the aircraft can land in about 500 m (1,640 ft). Using the powerful afterburning engine, take off can be made in an even shorter distance.

By late 1961 the Pratt & Whitney JT8D bypass engine had been chosen to power the aircraft. The engine was to

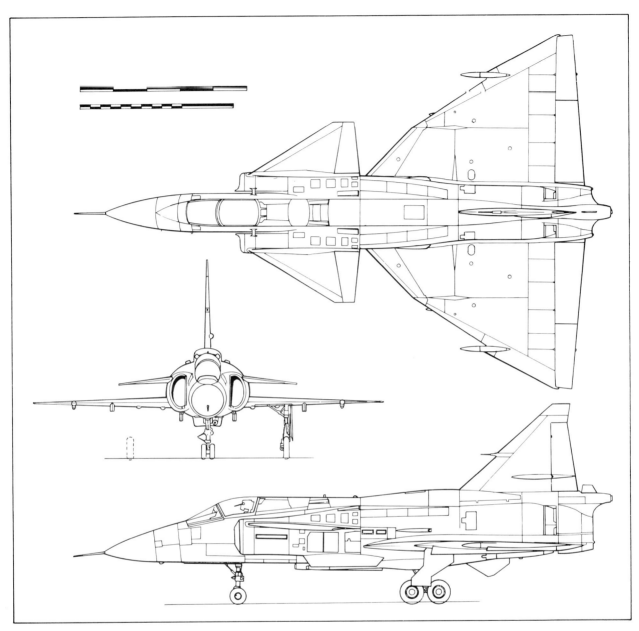

Saab JA 37 Viggen

be built under licence by Flygmotor (as the RM 8) and be further developed in Sweden for supersonic flight using a new afterburner. The basic engine was optimized for subsonic airline operations in aircraft such as the Douglas DC-9 and the Boeing 727. The RM 8, therefore, is a completely unique engine only produced in Sweden. For several years, it was one of the most powerful military jet engines in the world, with a static thrust in the first production version RM 8A of 11,800 kp (26,000 lb).

Yet another advanced technical solution chosen for the aircraft was the use of a central digital computer capable of controlling many functions, including navigation, sighting and weapon functions, and fuel monitoring, in order to reduce the workload of the pilot and to enable him to concentrate on tactical behaviour. An advanced automatic flight control system (AFC) can take over a larger (or smaller) part of the control to enable the pilot to analyse radar information.

A special microwave tactical instrument landing system (TILS) has also been installed in the aircraft and on the ground. In order to get the aircraft into small hangars and shelters, the vertical fin can be folded.

To control the System 37 technically as well as economically the 1965 Parliament decided that a special Project Directorate should be formed inside the Air Board to ensure that the System would meet both Air Force and Government requirements. The new Directorate was formed on 1 July, 1965, with Lars Brising, the former technical director of Saab as the head.

In April 1962 a new procurement system for System 37 was introduced by the Government with the appointment of Saab as main contractor. Associate contractors were also appointed with their responsibilities defined. Through the new contractual system, certain responsibilities for the system integration between the industrial companies involved were transferred to Saab while the Air Board maintained overall responsibility for the tactical and technical fulfilment of the specifications as well as the total economy of the system. To get a better control of the development work, a special industrial delegation (CB 37) headed by

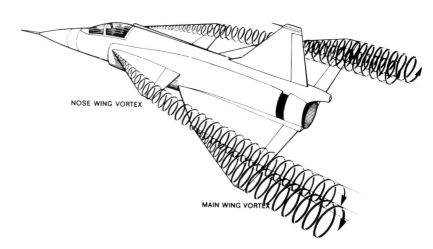

Viggen. Integration of nose wing vortex with main wing vortex. *(Saab)*

This photoraph of Viggen fore-plane and mainplane vortices does not conform to the pattern shown in the accompanying drawing. *(Flygvapnet/F 17)*

Harald Schröder and jointly staffed by the main and associated contractors was established at Saab. The PERT planning system (Project Evaluation and Review Technique) was introduced, and was used for control of more than 20,000 activities in the late 1960s.

In April 1965, a very detailed mock-up was completed for a first checking of the design drawings. It was also shown to a large number of invited guests including the Press. In the meantime the new aircraft had been given the name Viggen (The Thunderbolt) by the Commander-in-Chief of the Air Force General Lage Thunberg.

Preceded by extensive ground testing in Saab's new simulation centre, the first Viggen prototype made its first flight on 8 February, 1967, with Saab's chief test pilot Erik Dahlström at the controls. Everything went according to plan but as a result of ground testing the initial dihedral of the fore-plane

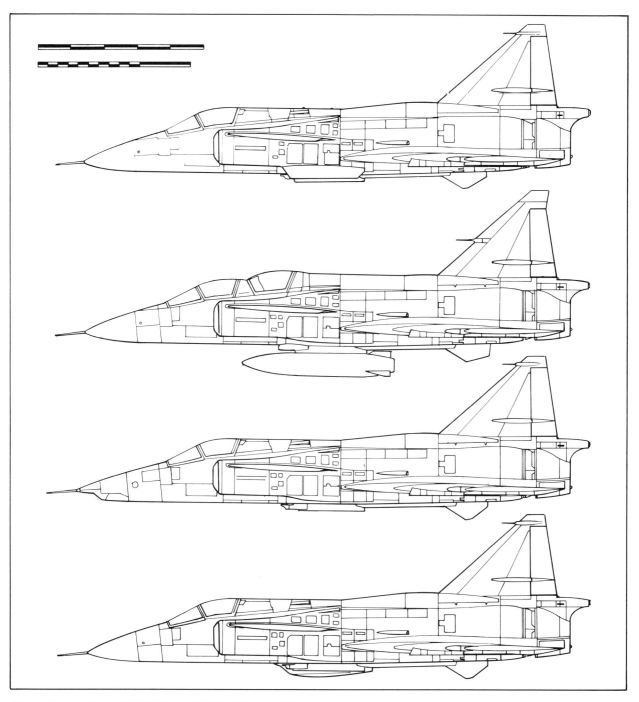

Top to Bottom: **Saab AJ 37 Viggen, SK 37, SF 37 and SH 37**

had been eliminated in the prototype shortly before the first flight. The first prototype was followed into the air by No. 2 in September 1967 and by No. 3 in March 1968.

The intensive flight testing revealed serious stability problems on the ground when using the thrust reverser. In addition, Viggen also developed certain aerodynamic problems with external stores. To compensate for the latter, a special dorsal bulge was introduced. The thrust reverser problems were also solved with the result that on the ground Viggen is today one of the most stable of jet fighters.

In April 1968 the Government authorized a first production order for 175 Viggen aircraft in the attack (AJ 37) and two-seat trainer version (Sk 37). The order was later amended to include an additional five AJ 37s and two reconnaissance versions, the SH 37 (sea surveillance) and SF 37 (photographic-reconnaissance).

The first production AJ 37s were delivered in 1971 with the production of the first Viggen generation continuing until 1979 with delivery of the 180th aircraft.

Most new aircraft invariably suffer from teething troubles, but in 1974–75 Viggen had more than its fair share. Three aircraft were lost in flight under unexplained circumstances, fortunately without loss of aircrew. After extensive investigations, it became clear that the main wing spar – a heavy light metal forging – had developed

The AJ 37 weaponry also includes the Rb 75 (Maverick) air-to-surface TV-guided missile. *(Saab/I. Thuresson)*

The AJ 37 was the first Viggen version to go into service. First delivery took place in mid-1971. *(Saab/W. Linder)*

microscopic cracks partly initiated by faulty drilling. Cracks could only be found in the wing spar design used in the first 28 production aircraft, after which another, much heavier, spar design had been introduced to allow for a longer service life. Contributing to the problem was that the high performance of the Viggen platform was used much more by the pilots than originally specified for the attack and reconnaissance versions. The heavier spar design was retrofitted in more than 20 aircraft. For later aircraft no modification was required.

Viggen is a fairly large aeroplane, its dimensions being mainly dictated by the large radar carried as well as by the bulky armament and reconnaissance equipment. The AJ 37 is capable of carrying up to three Rb 04E, the modernized version of this proven radar-homing, sea-skimming anti-ship

Viggen was one of the first aircraft in the world to be equipped with a central digital computer, considerably reducing the pilot's workload. *(Saab)*

The heavy podded rockets carried by Viggen have high precision. *(Saab)*

The AJ 37 with multiple launchers for sixteen 120 kg bombs. *(Saab)*

An AJ 37 Viggen prototype carrying two Rb 05A missiles and two Bofors rocket pods. *(Saab/I. Thuresson)*

missile. Other weapon alternatives include the Rb 05A radio-command guided air-to-surface missile developed by Saab and operational since 1972, the imported Rb 75 Maverick IR-homing air-to-surface missile, and the Rb 24 (Sidewinder) for self-defence. The AJ 37 can also carry two 30 mm Aden-type cannon in pods or sixteen 120 kg bombs in four multiple launchers. 80 kg flare bombs (Bofors-manufactured) are also part of the weapons inventory. In 1988 it was announced that the new RBS 15F will be added to the AJ 37's weaponry.

The long-range X-band monopulse radar in the AJ and SH 37 versions is of the PS-37/A type developed and produced by LM Ericsson (now Ericsson Radar Electronics), the central computer is the Saab CK-37, and the head-

A Viggen quintet taxi-ing on a narrow road during the brief period of sunshine in Arctic Sweden's winter days. Viggen is easily steered on narrow taxiways and roads. *(Saab/Å. Andersson)*

A JA 37 Viggen with one Sky Flash replaced by a camera pod for verification of missile performance. *(Saab/Å Andersson)*

up display British. The navigation system was originally planned to be of Decca type but was later replaced by a doppler-navigation system. An improved gyro system – FLI 37 – was developed by AGA (now Bofors Aerotronics), and a new air data system mainly using mechanical technology was developed by ARENCO. The autopilot was designed by Honeywell and produced under licence by Saab.

With the AJ 37 the semiconductor technology was introduced in virtually all avionics. The system was closely integrated with the central digital computer which received new high-capacity memories to handle the new tasks in the aircraft system including efficient autonomous tests of various functions.

The two reconnaissance versions

The AJ 37 Viggen with some of its armament alternatives: Three Rb 04E anti-ship missiles plus, near the aircraft *(left to right)*, Rb 28 (Falcon) air-to-air missiles, Rb 05A air-to-surface missiles, countermeasures pod, two pods each containing six 13.5 cm air-to-surface rockets and the Rb 24 (Sidewinder). The starboard side is identical except for a different type of countermeasures pod. The front row shows sixteen 120 kg bombs and a drop tank. *(Saab/I. Thuresson)*

and the two-seat trainer version are very similar to the attack version in platform equipment and performance.

The AJ 37 currently serves with the attack units based at Karlsborg (F 6), Såtenäs (F 7) and Söderhamn (F 15) and frequently participates in the regular airspace control missions normally flown by fighter units.

The SF 37 photographic-reconnaissance version made its first flight on 21 May, 1973, while production deliveries began in 1977. The SF 37 features advanced camera equipment for day and night photography (Sidewinders for self-defence) in the fuselage nose which houses four SKa 24 cameras (the same as used in the S 35E Draken) and two newly developed SKa 31 in addition to the VKa 702, and an IR camera.

The SH 37 is a combined reconnaissance and attack version. Its primary role is sea surveillance using a power-

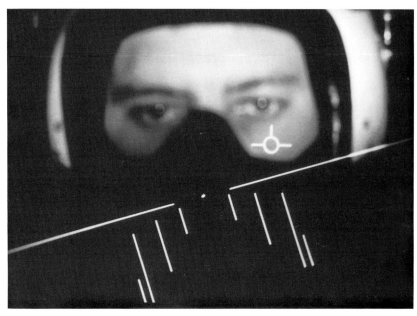

The head-up display in Viggen greatly facilitates the pilot's work. *(Saab/I. Thuresson)*

Viggen (AJ 37) cockpit is a roomy and efficient workplace. *(Saab)*

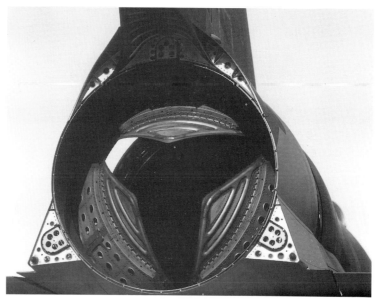

Viggen's main undercarriage is dimensioned for carrier-type landings. In recent examples the main undercarriage door is produced in composite materials. *(Saab/W. Linder)*

Close-up of the three titanium doors of Viggen's thrust reverser. About 60 percent of the non-afterburning thrust can be used for braking. *(Saab/Å. Andersson)*

The PS-37/A long-range radar in the AJ 37. It is somewhat modified in the SH 37 sea surveillance version. *(Ericsson)*

ful search radar as well as cameras carried in underwing pods for day and night use. The amount of information secured by the SH 37 is very great and the naval traffic around Sweden can easily be followed. The SH 37 can also carry certain of AJ 37's weapons. In the Viggen reconnaissance system is also included a mobile intelligence ground unit with advanced film development and interpretation facilities.

The two-seat Sk 37 trainer version can also be used for certain attack missions. This version first flew on 2 July, 1970. Production deliveries began in 1972 and all of these aircraft are currently based (1988) at the special conversion training unit at F 15.

Viggen was, of course, planned for the fighter role right from the beginning. Despite a generally good development potential of the aircraft's avionics

ABOVE: **Simple 'fishing rod' gears are used in turnaround areas when lifting missiles or other weapons into place on Viggen.** *(Saab)*

OPPOSITE TOP: **In the powerful RM 8 engine (Pratt & Whitney JT 8D produced under licence by Volvo Flygmotor) the afterburner is bigger than the basic engine.** *(Saab/Å. Andersson)*

OPPOSITE BOTTOM: **In 1971-72 Saab produced Viggen and Draken in parallel.** *(Saab/Å. Andersson)*

BELOW: **Viggen front fuselages undergoing equipment assembly before going to the final assembly line.** *(Saab/H. O. Arpfors)*

ABOVE: AJ 37s armed with 30 mm gun-pods. *(Saab)*

OPPOSITE TOP: The SF 37 is a specialized photographic reconnaissance version of Viggen in which a battery of cameras has replaced the radar. The reconnaissance equipment also includes an IR camera. Countermeasures, camera and flash pods are standard equipment as are Sidewinders. *(Flygvapnet/F 13)*

OPPOSITE BOTTOM: For conversion training two-seat Sk 37s are used. This version of Viggen is also used for certain attack missions. *(Saab/Å. Andersson)*

BELOW: The turnaround time for the JA 37 is less than 10 minutes. *(Flygvapnet/F13)*

A pair of Viggen reconnaissance aircraft flying low over the Baltic. *(Flygvapnet/F17)*

for the attack and reconnaissance roles, the extremely rapid avionics development during the 1960s and early 1970s clearly indicated that the avionics of the original multi-role platform would not meet the threat facing the fighter of the 1980s and 1990s. To fulfil the requirements specified for the JA 37 it would have to be equipped with completely new avionics (new pulse Doppler radar, new central computer, new head-up display, new autopilot, new air data unit, new inertial navigation, and a built-in cannon). The Swedish avionics development effort was mainly directed towards the Ericsson PS-46/A radar and Ericsson EP-12 cockpit display system, whereas the central computer, air data unit, autopilot and inertial navigation system were based on United States technology although mostly built under licence in Sweden (the computer and autopilot by Saab). The new IFF system was developed by Ericsson and the warning radar by SATT. The PS-46A radar alone was the result of a ten-year development programme. It features long range, look-down capabilities, track-while-scan and total digital mission data processing. It was, in fact, the first European-built airborne pulse Doppler radar.

The JA 37 fighter Viggen is virtually a new aircraft on the inside. A more powerful engine, a new pulse Doppler radar, a new central computer, a new 30 mm built-in cannon and new missiles are part of the system. The missiles are the British BAe Sky Flash (Rb 71) and the American AIM-9L Sidewinder (Rb 74). *(Saab/Å. Andersson and J. Dahlin)*

The ammunition tray for the Oerlikon cannon in the JA 37 is easily hoisted into position. The weapon has exceptional long-range performance. *(Flygvapnet/F 13)*

The SH 37 sea surveillance version of Viggen carries camera, flash and countermeasures pods as well as Rb 24 (Sidewinder) for self-protection. The SH 37 can also undertake attack missions. *(Flygvapnet/F 13)*

Saab 37 Viggen 161

The interface between the pilot and the aircraft is the basis for the efficiency of the Viggen fighter. Much development effort has gone into the co-ordination of computerized information displays, instruments and controls, with the result that the Viggen fighter is regarded as one of the best systems aircraft in service in the world.

The extensive use of the RM 8 engine in the AJ 37 and SH/SF 37 versions revealed certain limitations, especially functional reliability at high altitude. For this reason, a new RM 8 version had to be developed for the fighter version incorporating one additional fan stage in the low-pressure compressor. This engine, which is known as the RM 8B, has a static thrust with afterburner of 12,750 kp (28,100 lb) compared to 11,800 kp (26,000 lb) in the RM 8A.

The JA 37 is also equipped with a built-in 30 mm Oerlikon KCA cannon of exceptional performance. It fires 22

12,750 kp (28,000 lb) thrust at work; a Viggen afterner lit. *(Flygvapnet/F4)*

A full squadron of JA 37s ready to scramble at Norrköping. *(Flygvapnet/F13)*

Different colour schemes have been tested on the JA 37. Two different schemes are seen here on a trio of F 13 aeroplanes. *(Saab/Å. Andersson)*

rounds per second and features exceptionally high muzzle velocity (1,200 m/sec) making long-range firing (up to 2 km) possible. The complex aiming calculations are made in the JA 37 central computer. The HUD is used for aiming the gun. The JA 37 has also been integrated with new air-to-air missiles including the Rb 71 (BAe Skyflash) radar-homing weapon and the Rb 74 (Sidewinder AIM 9L). It can also carry attack weapons. The first JA 37 prototype made its first flight on 27 September, 1974, with Per Pellebergs at the controls. Six development aircraft were used for the JA 37 programme accumulating nearly 3,000 test flights to verify the system.

A unique feature of the Viggen (and other Swedish combat aircraft) is that it is designed to be made ready between missions mainly by National Servicemen. The turn-round time between Viggen fighter missions is less than 10 minutes. Each aircraft is serviced by five National Servicemen and there is only one chief mechanic for every two turn-round teams.

The first production aircraft flew on 4 November, 1977. Eventually a total of 149 JA 37s were ordered, with deliveries starting in 1979. At present (1988) five Wings, F 13, F 17, F 21, F 4 and F 16 operate the fighter version of the JA 37.

The JA 37 series brought Viggen production to a total of 329 aircraft.

The Ericsson PS-46/A pulse Doppler radar in a JA 37. The words beneath the triangle are 'Varning Rörlig Antenn' – Danger Active Antenna. *(Ericsson)*

AJ/SF/SH 37

Span 10.60 m (34 ft 9¼ in); length (including probe) 16.30 m (53 ft 5¾ in); height 5.60 m (18 ft 4½ in); wing area 52.20 sq m (562 sq ft). Loaded weight approximately 16,000 kg (35,274 lb). Maximum speed Mach 2+ at high altitude; cruising speed Mach 0.9; landing speed 220 km/h (137 mph); time to 11,000 m (36,000 ft) approximately 2 min; ceiling 18,000 m (59,000 ft); range 2,000 km+ (1,243+ miles).

JA 37

Span 10.60 m (34 ft 9¼ in); length 16.40 m (53 ft 9¾ in); height 5.90 m (19 ft 4¼ in); wing area 52.20 sq m (562 sq ft). Loaded weight approximately 18,000 kg (39,683 lb). Maximum speed Mach 2+ at high altitude; cruising speed Mach 0.9; landing speed 220 km/h (137 mph); ceiling 18,000 m (59,000 ft); range 2,000 km+ (1,243+ miles).

Saab 37 Viggen production serials

AJ 32 prototypes: 37001–37007
Sk 37 prototypes: 37800
AJ, SH, SF and JA: serials not available

Due to its special swept-forward wing configuration the crew of the MFI-17 has good visibility both above and below the wing. *(Saab/Å. Andersson)*

Saab MFI-15/17 Safari/Supporter

The MFI-17 Supporter has six hard-points under the wings and can carry up to 300 kg of external stores. *(Saab/Å. Andersson)*

In March 1968, Saab acquired Malmö Flygindustri AB (MFI), a light-aircraft manufacturer based at Malmö and originally formed in 1939 as AB Flygindustri at Halmstad. The acquisition included the MFI-15 trainer and army observation aircraft under development with Björn Andreasson as chief designer. The aircraft, named Safari, was planned mainly as a primary trainer for the Swedish Air Force and as a replacement for the Piper L-21B of the Swedish Army. The MFI-15 prototype made its first flight on 11 July, 1969. Although carefully matched to Swedish requirements the MFI-15 failed to impress the Defence Material Administration (FMV) which in 1969 had already selected the British Beagle Bulldog as an Air Force primary trainer. The Bulldog was also selected for the Army for technical commonality reasons, although it was later replaced by helicopters.

The first major customer for the Saab MFI-15/17 was Pakistan in 1974. It has also produced the aircraft under licence. Four Swedish-built examples are seen here. *(Saab/Å. Andersson)*

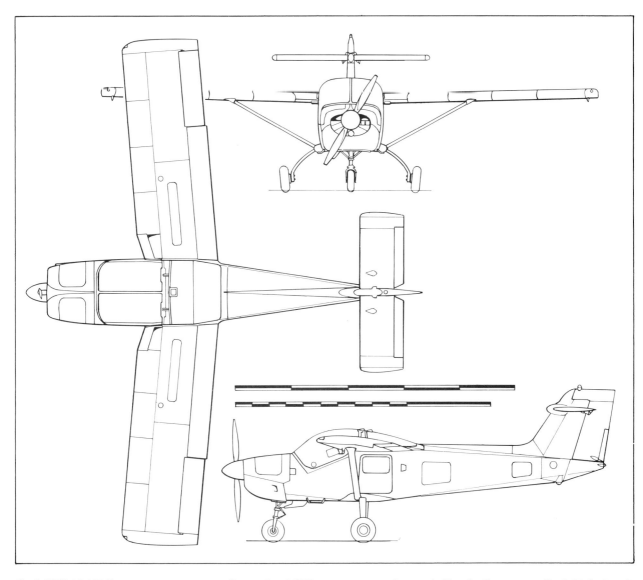

Saab MFI-15/17 Supporter

Saab was not too discouraged by the negative Swedish military decision and continued to develop the MFI-15 Safari into the MFI-17 Supporter version with six underwing hard-points for external stores. This version made its first flight in July 1972.

The MFI-17 is a 2/3-seat all-metal aircraft powered by a 200 hp Lycoming IO-360-A1B6 flat-four engine. The third seat (facing aft) can be reached from a door under the port wing. 300 kg of external stores can be carried.

Due to its unusual slightly forward-swept shoulder wing configuration, the aircraft provides excellent (helicopter-like) visibility in combination with

Gun pods of different types can be carried by the Supporter. *(Saab/J. Carlsson)*

A special relief operation in Ethiopia was undertaken in the late 1970s using a number of MFI-15s.

The clean instrument panel of the **MFI-15/17.** *(Saab/I. Thuresson)*

Denmark ordered thirty-two MFI-17 (T-17 in Denmark) in 1975, and in 1981 Norway ordered the first of a total of nineteen. *(Saab/Å. Andersson)*

outstanding STOL characteristics. The main undercarriage legs are of glass-fibre design.

The first country to order the MFI-15 was Sierra Leone which bought four. In 1974, Pakistan ordered a considerable number of aircraft, also acquiring a licence to produce it (under the designation Mushshaq). A year later Denmark ordered thirty-two MFI-17s (T-17 in Denmark) and in 1981 Norway ordered the first of nineteen aircraft. Approximately 300 MFI-15/17 aircraft have been produced.

MFI 15/17

Span 8.85 m (29 ft ½ in); length 7.00 m (22 ft 11½ in); height 2.60 m (8 ft 6⅓ in); wing area 11.90 sq m (128 sq ft). Empty weight 646 kg (1,423 lb); loaded weight 1,200 kg (2,645 lb). Maximum speed 236 km/h (147 mph); cruising speed 208 km/h (129 mph); landing speed 90 km/h (56 mph); initial rate of climb 5.5 m/sec (1,082 ft/min); ceiling 4,100 m (13,450 ft); range 1,050 km (653 miles).

Saab MFI 15/17 production serials

MFI 15 prototype: 15001
MFI 15/17:   15201–15209 Royal Danish Air Force
15210–15218 Royal Danish Army
15219–15232 Royal Danish Air Force
15236–15274 Pakistan Air Force
15803–15817, 15836–15840 Royal Norwegian Air Force
Other serials have not been released for publication.

HB-AHI, a Crossair Saab 340, was used during an extensive demonstration tour to the Far East in early 1985. Here it is seen over Hong Kong. *(Norman Peeling)*

Saab 340

The Saab 340 is normally configured to seat 35 passengers.

On 25 January, 1980, Saab-Scania AB of Sweden and Fairchild Industries Inc of the United States announced a joint decision to go ahead with the project definition phase for a twin propeller-turbine regional airliner seating about 30 passengers and with a cruising speed of some 480 km/h (300 mph). This historic decision had been preceded by several years of extensive project and market studies by both partners. In addition, Fairchild had established itself as a leading supplier of smaller, commuter-type aircraft seating around 20 passengers, and of executive aircraft, the Metro and Merlin. Saab, for its part, had worked on several different twin propeller-turbine projects. The high-wing project Saab 1084 for both commercial and military

use was almost ready for go-ahead in early 1979. However, the growing development and launching costs, as well as the strong desire to secure a larger home market, prompted Saab to look for a partner. As approximately half the world market for this class of aircraft was to be found in the United States, an American partner was the natural choice.

The contacts with Fairchild were intensified and in June 1979 a preliminary agreement was signed to conduct extensive joint project and market studies. By the end of 1979 about 50 engineers from each company were already working as a team, mainly at Fairchild Republic on Long Island. The different projects were re-aligned and transformed into one, a low-wing aeroplane exclusively designed for airline operation.

First unveiled at a Stockholm press conference on 25 January, 1980, the aircraft was going to use new technology in airframe and engine design in order to achieve superior operating economy.

A work-split between the two partners was made at an early stage under which Fairchild would be responsible for the design and manufacture of the wings together with engine nacelles and empennage, and with Saab taking responsibility for fuselage design and manufacture, systems integration, flight testing and certification. Legally, the aircraft was Swedish and to be initially certificated by the Swedish Board of Civil Aviation. The powerplant was to be certificated by the US authorities. Designated the Saab-Fairchild SF 340, the new aircraft was designed for 34 passengers in its initial configuration.

In June 1980 the General Electric CT7 engine of 1,700 shp was selected to power the SF 340. At the same time modern Dowty four-bladed propellers were specified. The propeller blades are of composite materials.

In the airframe structure, more bonding has been used than in any other airliner to give better fatigue life and better resistance to corrosion. A new, low-drag aerofoil has also been used. A major technical milestone was the decision to include a modern all-digital avionics system with autopilot and flight director as standard equipment. The system is similar to that used by Boeing in its 757 and 767 airliners and gives unique capabilities for a shorthaul propeller-turbine airliner.

During development work the passenger cabin, which is pressurized to maintain sea-level cabin altitude up to 12,000 ft (3,650 m) was slightly re-arranged to accommodate 35 passengers as standard. A special corporate

The Saab 340 offers a cruising speed of 285 knots and a fully loaded range of about 800 nautical miles. *(Saab/J. Dahlin)*

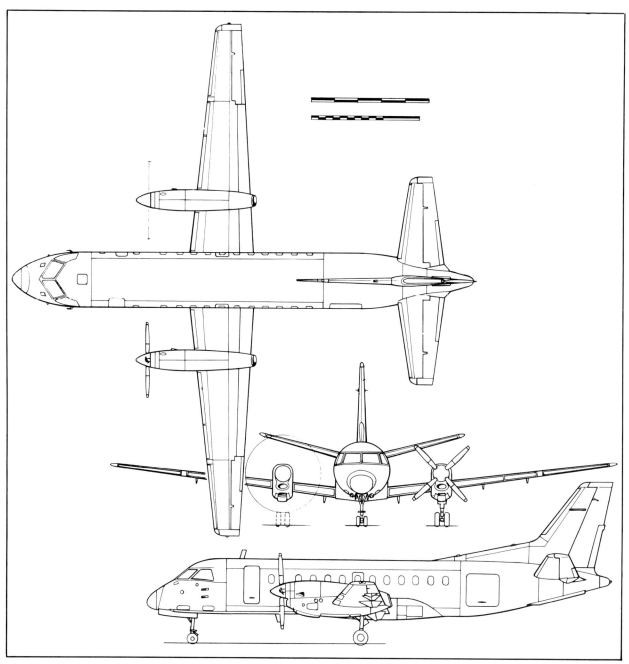

Saab 340

or executive version has also been developed intended mainly for the US market.

For the production and final assembly of the aircraft in Sweden, Saab-Scania in 1981–82 built a completely new factory with a floor space of 25,000 sq m (269,100 sq ft) involving an investment of more than 250 million Swedish Crowns. Later, additional production facilities were built. In 1980 Saab-Scania obtained from the Swedish Government a loan of 350 million Swedish Crowns to help finance development and production. The loan will be paid back in the form of royalties on each aeroplane sold.

The first prototype (SE-ISF) was rolled out of the Linköping factory on 27 October, 1982, and made its first flight on 25 January, 1983, three years to the day from signing the agreement with Fairchild. Saab test pilots Per Pellebergs and Erik Sjöberg were at the controls.

The SF 340 was unique from many points of view. It was not only the first airliner developed with partners on each side of the Atlantic, it was also the first airliner ever developed to meet the new joint European requirements (JAR 25) as well as the highest US airworthiness regulations (FAR 25).

After an intensive ground and flight-test programme involving five aircraft and 15 test pilots accumulating a total of 1,730 flying hours, the SF 340 was awarded its type certificate by the Swedish Board of Civil Aviation on 30 May, 1984. On 29 June, nine other European JAR member nations and the United States simultaneously signed airworthiness acceptance of the aircraft. In October, Australia became the 12th nation to approve the 340. This is a truly unique performance in civil aviation history.

The SF 340 went into scheduled commercial service on 14 June, 1984, with its launch customer, Crossair of Switzerland, which had demonstrated its confidence in the new aircraft by already ordering the SF 340 in November 1980, only two months after the Saab-Fairchild decision for a full go-ahead on a design, development, production and marketing programme.

In October 1984 the aircraft went into scheduled service in the United States with Comair of Cincinnati, Ohio, which one month later ordered a total of 15 aircraft.

In September 1984, a number of engine malfunctions occurred which led to the grounding of the aircraft by the authorities in Sweden, Switzerland and the United States. However, the problems were rapidly solved by the engine manufacturer, and the aircraft quickly returned to revenue service.

In October 1985, it was jointly announced by Saab and Fairchild that

By the end of 1988 approximately 130 Saab 340 twin propeller-turbine regional airliners had been delivered to customers. In late 1985, Saab-Scania took complete responsibility for the 340 programme. *(Saab/J. Lindahl)*

from 1 November Saab would assume overall control of the SF 340 programme, with Fairchild continuing as a sub-contractor until 1987, *i.e.* up to and included aeroplane No. 109. By the autumn of 1987 wing and empennage production had been transferred to new facilities at Linköping. Meanwhile the marketing of the aircraft proceeded successfully and in June 1986, fifty aircraft were in service with seven airlines and two corporate customers on three continents.

In September 1987 the 100th aircraft was delivered, to Salair of Sweden. The largest fleet of the Saab 340 – as the aircraft was redesignated – is that ordered by Crossair, with twenty-four. Other major fleet owners are Comair (16) Swedair (11) and Dallas-based Metroflight (16).

By the end of 1988 there were twenty-six operators of the Saab 340; eight in the United States, thirteen in Europe, one in Australia, three in Latin America and one in Taiwan. More than 130 aircraft were in service. More than 12 million passengers had flown in the aircraft in approximately 750,000 flights.

More metal-to-metal bonding is used in the Saab 340 than in any other airliner in order to get a better fatigue life and resistance to corrosion. *(Saab/J. Lindahl)*

In June 1985, Saab confirmed its lead over the competion by demonstrating two 340s at the Paris Air Show. *(Saab/J. Dahlin)*

LEFT: **The Saab 340 was the first regional airliner to feature a completely digital cockpit display system as standard equipment.** *(Saab/N. G. Widh)*

In October 1984 the Saab 340 went into scheduled service in the United States with Comair of Cincinnati, Ohio. *(Saab/Å. Andersson)*

A Saab 340 of Norving Norway. *(Norman Peeling)*

Technically, the aircraft has undergone evolutionary development during its four-year production life, aimed mainly at improving economy, reliability, versatility and performance. In May 1985, the aircraft was certificated in a more powerful configuration with two 1,735 shp CT7-5A2s and larger diameter Dowty propellers permitting a higher take-off weight of 12,372 kg (27,275 lb) compared to the original take-off weight of 11,794 kg (26,000 lb). A quick-change (QC) version for passengers or cargo has also been developed.

On 9 September, 1987 - on the occasion of the delivery of the 100th aircraft - Saab announced a new and more powerful B version mainly intended for 'hot and high' conditions. The higher performance will be possible through installation of General Electric CT7-9B engines delivering a maximum take-off rating of 1,870 shp. The higher power will provide better altitude performance, including faster climb and a cruising speed of 285 knots (272 knots for the 1,735 shp engines). The Saab 340B will allow a maximum take-off weight of 28,500 lb and provide an extension of fully loaded range to about 800 nautical miles. Furthermore the new version will introduce enhanced centre-of-gravity range permitting higher payloads with emphasis on baggage and freight. A new brushless environmental control system fan will further reduce the internal noise level. The Saab 340B is scheduled for delivery to customers during 1989. Retro-fitting of existing aircraft to B standard is possible. In early September 1988 Crossair again became the launch customer with a firm order for five 340Bs.

After the early, mainly engine-oriented teething troubles, the Saab 340 is now fully recognized by the market as a reliable and profitable airliner. Its dispatch reliability is claimed to be as high as 99 percent. Saab has thus also amply demonstrated its capabilities in the highly competitive commerical aircraft market and did this over the short period of four years.

Saab 340A

Span 21.44 m (70 ft 4 in); length 19.72 m (64 ft 8 in); height 6.86 m (22 ft 6 in); wing area 41.81 sq m (450 sq ft). Operating empty weight 7,899 kg (17,415 lb); maximum take-off weight 12,372 kg (27,275 lb). Maximum cruising speed (at 26,000 lb all-up weight) 504 km/h (313 mph); best-range cruising speed (at 26,000 lb) 463 km/h (288 mph); landing speed 200 km/h (124 mph); maximum rate of climb at maximum take-off weight 9.13 m/sec (1,800 ft/min); service ceiling 7,620 m (25,000 ft); range with 35 passengers and reserves 1,553 km (965 miles); take-off and landing field length (FAR 25) 1,220 m (4,000 ft).

Saab 340B

Span 21.44 m (70 ft 4in); length 19.73 m (64 ft 9 in); height 6.87 m (22 ft 6 in); wing area 41.81 sq m (450 sq ft). Operating empty weight 8.036 kg (17,715 lb); maximum take-off weight 12,930 kg (28,500 lb). Maximum cruising speed (at 26,000 lb all-up-weight) 531 km/h (330 mph); best-range cruising speed (at 26,000 lb) 463 km/h (288 mph); landing speed (at maximum landing weight) 212 km/h (132 mph), maximum rate of climb at maximum take-off weight 10.16 m/sec (2,000 ft/min); service ceiling 7,620 m (25,000 ft); range with 35 passengers and reserves 1,553 km (965 miles).

Saab 340A production (to 29 August 1988)

001	SE-1SF	Prototype
002	SE-1SA	Prototype Converted to 340B
003	SE-1SB	Pre-production, to Fairchild Industries as N9668N
004	N 340CA	Comair
005	HB-AHA	Crossair
006	N360CA	Comair
007	HB-AHB	Crossair
008	G-BSFI	Birmingham Executive, to Manx Airlines G-HOPP
009	HB-AHC	Crossair
010	N 370CA	Comair
011	N 342AM	Air Midwest
012	N 380CA	Comair
013	SE-1S0	Swedair
014	N 340SF	Fairchild, Mellon Bank, Amcomp, Comair
015	SE-1SP	Swedair
016	VH-KDK	Kendell Airlines
017	SE-1SR	Swedair
018	HB-AHD	Crossair
019	N 343AM	Air Midwest
020	HB-AHE	Crossair
021	N 341CA	Comair
022	N 19M	340 Associates
023	N 342CA	Comair
024	N 343CA	Comair
025	N 344CA	Comair
026	HB-AHF	Crossair
027	N 320PX	Republic Express, Northwest Airlink
028	N 347CA	Comair
029	N 100PM	Philip Morris
030	N 344AM	Air Midwest
031	N 321PX	Republic Express, Northwest Airlink
032	N 346AM	Air Midwest
033	SE-1SS	Swedair
034	N 356CA	Comair
035	SE-1ST	Swedair
036	N 200PM	Philip Morris
037	LN-NVD	Norving, to Crossair
038	HB-AHG	Crossair
039	N 347AM	Air Midwest
040	HB-AHH	Crossair
041	N 322PX	Republic Express, Northwest Airlink
042	SE-1SU	Swedair
043	HB-AH1	Crossair
044	N 357CA	Comair
045	SE-1SV	Swedair
046	N 323PX	Republic Express, Northwest Airlink
047	N 358CA	Comair
048	N 324PX	Republic Express, Northwest Airlink
049	HB-AHK	Crossair
050	N 340SA	Saab Aircraft of America, Kelly Springfield N 44KS
051	N 325PX	Republic Express, Northwest Airlink
052	VH-KDP	Kendell Airlines

Republic Express (now Business Express) is another major operator of Saab 340s in the United States. *(Norman Peeling)*

An unusual sight, ten of Crossair's Saab 340s seen together, at Basle-Mulhouse Airport. *(Saab)*

053	N 359CA	Comair	073	N 935MA	Air Midwest	100	SE-ISK	Salair *Blaaklinten*
054	N 326PX	Republic Express, Northwest Airlink	074	N 406BH	Eastern Express	101	N 343BE	Business Express
			075	D-CDIB	Delta Air	102	N 365MA	Metroflight
055	LN-NVE	Norving, leased as PH-KJH to Netherlines	076	N 329PX	Northwest Airlink	103	N 365MA	Metroflight
			077	N 922MA	Air Midwest	104	N 344BE	Business Express
056	N 361CA	Comair	078	N 407BH	Eastern Express	105	N 744BA	Brockway Air
057	N 420BH	Bar Harbor Airlines (Eastern Express)	079	N 340PX	Northwest Airlink	106	LV-AXW	L.A.E.R. (Linea Aereas Entre Rios)
			080	SE-1SY	Swedair			
058	N 402BH	Bar Harbor Airlines (Eastern Express)	081	F-GELG	Europe Aero Service	107	N 367MA	Metroflight
			082	HB-AHL	Crossair	108	N 345BE	Business Express
059	N 327PX	Republic Express, Northwest Airlink	083	F-GFBZ	Britair	109	N 368MA	Metroflight
			084	HB-AHM	Crossair	110	N 369MA	Metroflight
060	N 403BH	Bar Harbor Airlines (Eastern Express)	084	HB-AHM	Crossair	111	N 745BA	Brockway Air
			085	F-GGBJ	Air Limousin	112	N 370MA	Metroflight
061	N 404BH	Bar Harbor Airlines (Eastern Express)	086	F-GBBV	Air Limousin	113	HB-AHO	Crossair
			087	LN-NVF	Norving, to Swedair	114	N 371MA	Metroflight
062	N 340BE	Business Express	088	HB-AHN	Crossair	115	N 372MA	Metroflight
063	N 341BE	Business Express	089	N 360MA	Metroflight	116	D-CDIC	Delta Air
064	N 320CA	Comair	090	N 740BA	Brockway Air	117	F-GHDB	Britair
065	OH-FAA	Finnaviation	091	N 361MA	Metroflight	118	N 373MA	Metroflight
066	OH-FAB	Finnaviation	092	N 742BA	Brockway Air	119	N 373MA	Metroflight
067	SE-1SX	Swedair	093	N 743BA	Brockway Air	120	HB-AHP	Creossair
068	N 328PX	Northwest Airlink	094	LV-AXV	Transportes Aereos Neuquen	121	N 109TA	Tempelhof Airways
069	N 691P	LAPA (Lineas Aereas Privadas Argentinas)				122	HB-AHQ	Crossair
			095	N 362MA	Metroflight	123	N 3Y5MA	Metroflight
070	OH-FAC	Finnaviation	096	N 342BA	Brockway Air	124	D-CDID	Delta Air
071	D-CD1A	Delta Air	097	SE-1SZ	Swedair	125	N 125CH	Chautauqua Airlines
072	N 72LP	LAPA (Lineas Aereas Privadas Argentinas)	098	N 363MA	Metroflight	126	HB-AHR	Crossair
			099	N 364MA	Metroflight	127	B-12200	Formosa Airlines

About 30 percent of Gripen's structure is made of composite materials. *(Saab)*

Saab 39 Gripen (The Griffin)

An artist's impression of Gripen's cockpit with its Ericsson EP-17 display system. *(Ericsson Radar Systems)*

Flexibility is the keyword for the JAS 39 Gripen multi-role combat aircraft which is intended to replace, starting in the early 1990s, all versions of Viggen. The basic idea is that each individual aircraft should be able to undertake either fighter, attack or reconnaissance missions, making it difficult for an aggressor to assess exactly the combat potential of the Swedish Air Force units. The cockpit design has therefore received great attention and incorporates the most modern display and control technology to reduce the pilot's workload. Gripen is a light aircraft – its take-off weight is approximately 8,000 kg (17,637 lb) – and it is the first light combat aircraft to make full use of the advanced technology now available in engines, in structural materials, fly-by-wire control systems and micro-electronics. Fighter requirements have dictated the aircraft's performance, which means very high speed, acceleration and turning capability. This is in contrast to Viggen where the attack requirements largely dictated performance. The attack version of Viggen, the AJ 37, was also the first to go into service.

As a fighter Gripen will have full all-weather, all-altitude air defence capability, being armed with: a) the current generation of infra-red and radar homing air-to-air missiles; b) the new generations of air-to-air missiles for medium ranges and close-in fighting and c) a built-in, high performance cannon.

The target search and acquisition system in the air defence mission is a powerful PS-05/A Pulse Doppler lookdown/shoot-down radar developed by Ericsson. It is the first radar in Western Europe for a light combat aircraft having complete operating modes for fighter, attack and reconnaissance missions, and optimized to work under enemy countermeasures. Most functions are controlled in the software of a fully programmable signal and data processor using the Ericsson SDS 80 standardized computing system. The radar uses FM pulse compression and a large number of frequencies and waveforms as well as having frequency agility.

The operating functions in the radar system for fighter missions are: a) target search and tracking of several targets at long ranges; b) wide-angle, quick-scanning and lock-on at short ranges as well as c) fire control for missiles and cannon. The multiple-target radar function will provide Gripen with increased fire power compared to existing fighters.

For attack/reconnaissance the radar operating functions are: a) search against sea and ground targets; b) mapping with normal and high resolution; control for missiles and other attack weapons and d) obstacle avoidance and navigation. The complete target search and acquisition system is designed to include two separate parts; the nose-mounted radar and pod-mounted infra-red equipment (FLIR). The aircraft will also have excellent capability for autonomous air defence, i.e. without ground control support, thanks to its long radar range and good fuel economy.

In the attack role, Gripen is designed to carry heavy external loads including: a) electro-optically guided missiles and bombs; b) area weapons and c) anti-ship missiles. It will also carry very advanced countermeasures equipment, both built-in and external.

Weapon alternatives announced for Gripen are:
- One internal 27 mm Mauser BK 27

The General Electric/Volvo Flygmotor RM 12 (F404J) turbofan which powers Gripen. With afterburning it is rated at 80.5 kN (18,100 lb). The engine installation is easily accessible. *(Saab)*

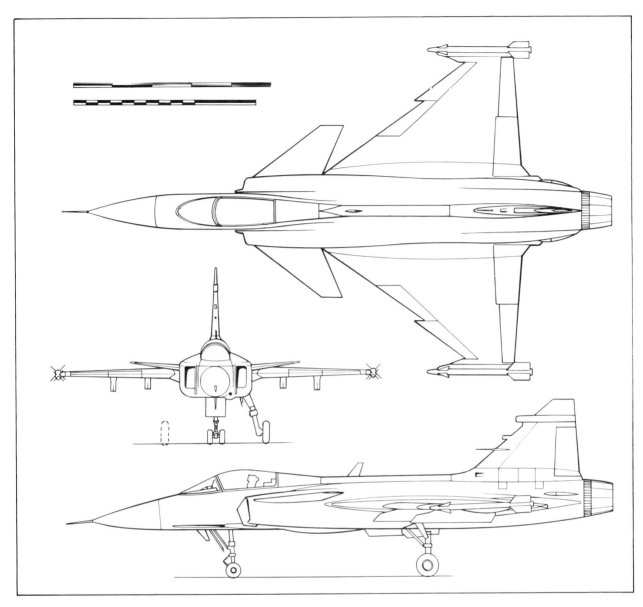

Saab 39 Gripen

automatic cannon (capable of firing 1,700 rounds per minute)
- Rb 74 (Sidewinder AIM-9L) air-to-air infra-red-homing missiles
- Rb 71 (Skyflash) radar-homing air-to-air missiles
- Rb 71A (A Skyflash development)
- Rb 75 (Maverick) air-to-surface missile
- Rb 15F anti-ship missile
- MBB area weapons

Six hard-points for weapons are carried.

The low weight and size of Gripen was made possible through extensive use of new technology. Of decisive importance in aircraft design has been the tremendous engine development during the past 15 years. For a given thrust, engine weights have been reduced by 50 percent and the number of parts by 30 percent while fuel consumption has been significantly reduced.

The General Electric/Volvo Flygmotor RM 12 (F404J) turbofan selected to power Gripen is rated at 53.4 kN (12,000 lb) dry and 80.5 kN (18,100 lb) with afterburning. Compared to US twin-engine installations, the RM 12 is modified for higher flight safety in the event of bird strikes and adaptation to Swedish operational profiles.

Sweden's extensive experience of the canard configuration used in Viggen shows that this is the most efficient for combining good high-speed performance with good low-speed and landing characteristics. The most important difference between Viggen and Gripen is that the latter's canard foreplane will be used as normal control surfaces. This means that there are control surfaces both forward and aft of the aircraft's centre of gravity, which gives enhanced manouevrability as well as reduced drag.

Since every Gripen will perform air defence, attack and reconnaissance missions, great demands are placed on the flight-control system to accommodate different loads and flight conditions.

Gripen is equipped with a Lear Siegler triple redundant electrical control system (fly-by-wire) including automatic flight control functions. The it includes the Flight Data Display for major flight and tactical information, the Electronic Map Display showing tactical information, geographical features and obstacles.

The Multisensor Display shows radar information and TV/IR imagery. The Head-up Display presents all vital information in the pilot's line of sight using advanced diffraction optics to **provide a wide field of view and high** (Saab-Scania, Volvo-Flygmotor, Ericsson Radio Systems and FFV Maintenance) on 3 June, 1981. A formal contract between the industry group and FMV was signed on 30 June, 1982, covering five flying prototypes and an initial production batch of thirty aircraft. The contract also includes an option for delivery of a further 110 production aircraft by the year 2000.

The development programme in-

The Saab aircraft product programme for the 21st Century; the JAS 39 Gripen and the Saab 340 regional airliner.

digital Gripen technology provides great development potential through flexible use of control methods and flight paths. One result is more efficient use of the cannon against both air and ground targets.

Structurally, Gripen will represent something of a revolution for the Swedish aircraft industry as almost 30 percent of the structure is made of composite materials (carbon-fibre reinforced plastics) giving a 25 percent weight saving for a given strength. The first prototype carbon-fibre wings were manufactured by British Aerospace but quantity production will be handled by Saab. The Gripen cockpit contains an advanced Ericsson EP-17 electronic display system using three head-down displays and one head-up display. The EP-17 uses two SDS 80 computer systems for multi-mode use and future development flexibility, and brightness image. Conventional instruments are only used as back-ups.

The computer hardware in Gripen uses modern large-scale, integrated circuit technology and high-density packaging for very small size and high performance. The total computing power of Gripen will be more than five times greater than in the JA 37 within almost the same hardware volume. It carries a total of nearly 30 computers.

Other important avionics in Gripen include the Bofors Aerotronics AMR 345 VHF/UHF AM/FM communications transceiver and the Honeywell laser inertial navigation system.

Swedish Government approval for development and production of the Gripen programme was given on 6 May, 1982, on the basis of an evaluation by the Defence Material Administration (FMV) of an extensive offer submitted by the JAS Industry Group cludes extensive ground testing as well as flight testing of avionics in several Viggen development support aircraft. The first such aircraft was a JA 37 first flown on 14 September, 1982, equipped with a fly-by-wire flight control system. A full-scale mock-up of Gripen was shown to the Press in February 1986 and on 26 April, 1987, the first prototype was unveiled. The first flight was expected to take place before the end of in 1988. This represents a delay of more than a year compared to the original time schedule, mainly due to delays in subsystems development, notably the flight control system, but Saab-Scania is confident that the Service introduction will remain on schedule, i.e. early 1992.

More than 3,000 people in the Swedish aircraft industry including 2,000 at Saab-Scania have participated in the development of the JAS 39 Gripen

The official roll-out of the JAS 39 Gripen light multi-role combat aircraft took place on 26 April, 1987. Gripen is the first light combat aircraft using new technology throughout. *(Saab/Å. Andersson)*

system. The Saab Aircraft Division is responsible for the development and manufacture of the basic aircraft and overall systems integration. Project manager at Saab is Tommy Ivarsson.

Gripen is the sixth generation of jet combat aircraft developed in Sweden.

Saab 39 Gripen

Span 8.00 m (26 ft 3 in); length 14 m (45 ft 11¼ in). Take-off weight 8,000 kg (17,637 lb). Maximum speed supersonic at all altitudes. No other data and performance figures have been released. Figures are approximate.

Footnote: This aircraft first flew on 9 December, 1988.

One of F 13's J 22 fighters. *(Flygvapnet/F 13)*

Rotating jigs were used for J 22 wing assembly. 198 aircraft were built. *(Text & Bilder)*

Appendix

FFVS J 22

Although in no way a Saab product, the FFVS J 22 'stop gap' fighter represents an important chapter in the Second World War history of the Swedish aircraft industry. Its origin has already been discussed in the Historical Survey but more technical information is certainly justified.

The aircraft's chief designer, Bo Lundberg, completed his initial project study in October 1940 after his return from the United States where he had been posted to Vultee to oversee the manufacture of P-48 Vanguard fighters for the Swedish Air Force (never delivered because of the US arms export embargo in mid-1940). Lundberg and Colonel Nils Söderberg, head of the Air Material Department of the Air Board and chief architect of the J 22 programme, had agreed that in order to make the programme feasible the aircraft had to be designed by a team outside the already overburdened aircraft industry (Saab), using a large number of sub-contractors and prefer-

The FFVS J 22 was developed as a 'stop-gap' fighter in 1941-1942. *(Flygvapnet/F3)*

ably to be built in materials not in critical demand by the established aircraft industry. At an early stage the designers decided on steel and wood as the most suitable materials. The basic structure of the fuselage and wings were to be of welded stainless steel covered by load-carrying wooden panels according to a new method.

In view of the limited engine power available – the SFA-built copy of the 1,065 hp Pratt & Whitney Twin Wasp radial engine, great care had to be taken over the aerodynamic design of the P 22, as the project was initially designated. The fuselage was given the smallest possible cross-section aft of the engine and the form was made as nearly perfectly streamlined as possible without using double curvature (ruled out for production reasons). The

Warm weather servicing of a J 22. *(Flygvapnet/F 18)*

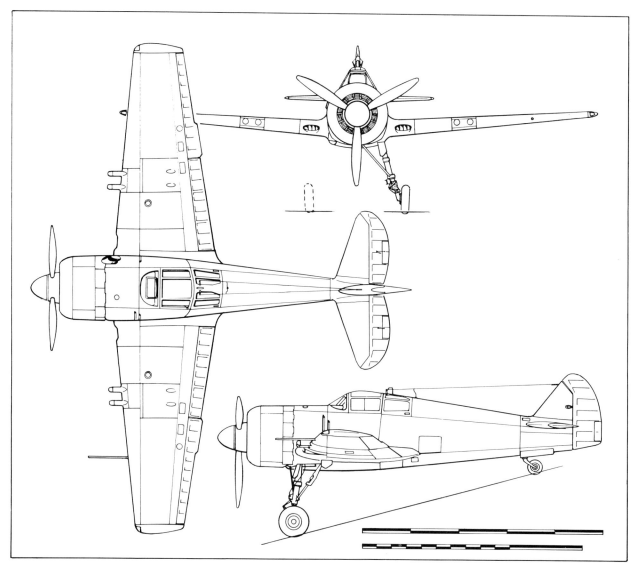

FFVS J 22

wing was designed to be fairly thin at the root but rather thick at the tips to prevent tip stalling.

The retractable undercarriage design was unusual in many respects. It was placed well forward and equipped with a mechanism to enable the doors to be closed when in the down position so as to reduce drag at take off and to prevent dirt from being flung into the wheel wells during taxi-ing, taking off and landing. Concrete runways were mostly a thing of the future. The aircraft was even designed with a retractable ski undercarriage but this was never produced in quantity owing to the availability of improved snow-clearing equipment.

Details of the project were settled by the end of 1940, and on 21 February, 1941, the Government approved the necessary funding for design and development work (including a mock-up) as well as two prototypes. The prototypes were to be built using the facilities available at the Aeronautical Research Institute (FFA) at Bromma Airport, Stockholm.

On 3 September, 1942, the first prototype had progressed so far that it could be weighed in order to establish the centre of gravity. As it transpired the dry weight was lower than calculated, 1,902 kg (4,193 lb) against the calculated 1,952 kg (4,303 lb): no mean performance by a fresh design team!

On 20 September the first flight took place with Major O. Enderlein at the controls. The flight lasted 40 minutes and was described as completely successful. Enderlein said the aircraft was 'easy to fly with excellent manoeuvrability. No dangerous tendencies were discovered despite the fairly high wing loading. The aircraft was easy to take off and land'. Subsequent flight testing, however, lead to a slight enlargement of the rudder. The landing run, using hard braking, was only 250 m (820 ft).

Built basically of steel and wood, the J 22 was produced using hundreds of suppliers outside the existing aircraft industry. *(Flygvapnet/F 10)*

The J 22 was also used for reconnaissance. Note cartridge collector box. *(Flygvapnet/F 3)*

As early as 6 October a diving test to 620 km/h (385 mph) indicated air speed was made.

In the meantime the structure was tested on the ground, with the wing spar successfully surviving a stipulated 180 percent safe design load without major deformations to the wooden panels.

The second prototype made its first flight on 11 June, 1943. However during the following two months the J 22 programme suffered a major setback through the loss of both prototypes. In the first accident, the test pilot, Lt B. J. E. Salwén, lost his life for reasons unknown. Indications were that he lost consciousness due to lack of oxygen.

After the second accident, the test pilot was able to walk away although he was considerably shaken. The engine had failed at flagpole height during a critical landing manoeuvre and the aircraft was a total loss.

But flight testing had already shown that there was nothing basically wrong with the aircraft; in fact the contrary was the case, and there was certainly no occasion for lengthy contemplations in those days. In fact, on 21 March, 1942, exactly six months before the first flight, an initial sixty aircraft had been ordered off the drawing-board. The new production machinery went into operation, with hundreds of suppliers participating in the programme. Some deserve special mention. The fuselage inner structure was produced by Hägglund & Söner at Örnsköldsvik whilst the wing inner structure was the responsibility of See Fabriks AB at Sandviken which together with AGA at Lidingö near Stockholm welded and hardened the wing spars and the other interior components. NK at Nyköping produced the wooden panels for the fuselage and the wings. Uno Särnmark at Gothenburg produced much of the electrical equipment and Nordiska Armaturfabriken (NAF) at Linköping most of the instrumentation.

Despite the loss of the two prototypes which delayed flight testing until the first production aircraft became available, deliveries to the F 9 fighter Wing at Gothenburg began on 23 October, 1943.

Eventually a total of 198 production aricraft were ordered. All but 18 of these were completed before a metalworkers union strike stopped production in February 1945. On 1 July, 1945, the new Air Force maintenance workshops at Arboga (CVA) were ready and their first job was the assembly of the last 18 aircraft which were allocated to a reconnaissance Wing, and equipped with an SKa 4 camera and therefore redesignated S 22. Two fighter versions were produced: 142 J 22As and 57 J 22Bs. The main difference was in the armament. The J 22A had two 8 mm and two 13.2 mm guns, the J 22B four 13.2 mm.

In performance the J 22 compared well with foreign fighters of the early 1940s. For example, it could outclimb the North American P-51D Mustang at altitudes below 4,500 m (15,000 ft) despite the limited power available. At higher altitude, the Mustang's supercharged Merlin engine turned the balance. It is interesting to note that the development costs including prototypes, testing and production tooling was only some 22 million Swedish Crowns compared to the total Air Force budget in 1942 of 213 million.

The J 22 programme meant that in 1943-44 the capacity of the Swedish aircraft industry could suddenly be increased by 40 percent. During 1944 a monthly production of 22 aircraft was achieved. The aircraft unit cost (excluding engine and instrumentation) was only 150,000 Swedish Crowns whereas the complete aircraft with engine, radio and instrumentation cost some 300,000. The total unit cost was affected by the fact that less than 600 Swedish Twin Wasps were produced.

FFVS J 22

Span 10.0 m (32 ft 9½in); length 7.8 m (25 ft 7 in); height 3.6 m (11 ft 10 in); wing area 16 sq m (172 sq ft). Empty weight 2,020 kg (4,453 lb); loaded weight 2,835 kg (6,250 lb). Maximum speed 575 km/h (357 mph); curising speed 340 km/h (211 mph); landing speed 140 km/h (87 mph); initial rate of climb 8.4 m/sec (1,650 ft/min); ceiling 9,300 m (30,510 ft); range 1,270 km (790 miles).

FFVS J 22 production serials

J 22 prototypes: 22001, 22002
J 22A and B: 22101-22298

Index

A 20 *see* Saab 37 Viggen *under* Saab aircraft
A 21 *see* Saab 21 *under* Saab aircraft
A 32 *see* Saab 32 Lansen *under* Saab aircraft
ABA *see* Swedish Air Lines
Acro Delta aerobatic team 129
Aeronautical Research Institute of Sweden (FFA) 11, 76, 184
Aeroplanvarvet Skåne (AVIS) 9
Aerotransport, AB (ABA) *see* Swedish Air Lines
Aerovias Brasil *88*, 90 92
AFF *see* Förenade Flygverkstäder
AFV *see* Flygplanverken, AB
AGA Company 42, 151, 185
Air Force Board
  civil aircraft 63, 90
  contract negotiations 18, 20, 30, 50, 54-5, 75, 104, 143
  farsightedness 39, 131-3
  responsibilities 42, 145-6
  wartime problems 22, 23, 24
  see also Basic Agreements
Air Force, Swedish
  aircraft orders *12*, 62, 68, 96, *100*, 126
  aircraft requirements 59, 66, 104, 177-18, 136
  formation 11, 12
  programmes 14-15, 36, 90
  strengths 14-15, 31, 39, 54
  tactical measures 27-8, *46*, 125
  wartime plight 20, 24
  see also F Wings
Air France 96
Air Material Department (AMD) 50
Airborne Mapping Ltd 71
Airline Pilot School, Netherlands 96, *101*
AJ 37 *see* Saab 37 Viggen *under* Saab aircraft
Albatros trainer 9
Amundson, Gen K. A. B. 13
Andersson, A. J. 11, 30, 75-6, 93, *95*
Andreasson, Björn 163
Antarctic Expedition (1951) 98
Arboga (CVA) 30, 40-1, 186
ARENCO Company 151
Armstrong Siddeley engines, Jaguar 11, *12*
Army, Swedish, aircraft and missiles 40, 59, 163
Ars, AB 18
Asea Company 30
ASJ *see* Svenska Järnvägsverkstäderna
ASJA *see* Svenska Järnvägsverkstädernas Aeroplanavdelning
Australia, aircraft orders 54, 171, 173
Austrian aircraft orders
  Draken 45, 53-4, 135-6, *141*
  Saab 29: *131*, 114-15, *115*
  Saab 105: *49*, 49, 53-4
  Safir *93*, 97
  other 63, *140*, 141, 142

AVIA Company 63
Avro Lancaster 39

B3LA 54, *55*
B 3 (Ju 86K) 12, 19, 22, *23, 24*, 68
B 4 (Hart) *13*, 14, *16*, 20, 22, 27
B 5 (8A-5) 19, *21*, 21, 22, 27
B 6 (2P-A) 24
B 16 (Ca.313) 24
B 17 *see* Saab 17 *under* Saab aircraft
B 18 *see* Saab 18 *under* Saab aircraft
Basic Agreements *(Ramavtal)*
  (1940): 24-6, 30, 31, 75
  (1949): 36
  (1961): 45
Beagle Bulldog 163
Belgium, NATO procurement *52*, 54
Bjurströmer, Bror 66, 87, *91*
Blériot monoplane (Thulin A) *8*, 9
Blomberg, Sven 11, 13
BOAC 92
Board of Civil Aviation (BCA) 58, 88, 90, 169, 171
Boeing aircraft
  B-17: *33*, 35
  Boeing 727: 145
  Stratocruiser 92
Bofors Aerotronics 151
Bofors Company 11, 15-16, 18, 31
  missiles 47, 48
  see also Saab-Bofors Missile Corporation
Bolinder-Munktell Company 31
Bosch 30
Bråsjö, Arthur 116
Bratt, Erik *42, 125,* 125
Brazil, aircraft orders *88*, 90
Brewster Buffalo 22
Brising, Lars *42, 52,* 104, *105,* 145
Bristol Aeroplane Company 11
Bristol engines
  Jupiter 11, *12*
  Mercury 11, *16,* 26, *59,* 59, 61
  Taurus 20, *22,* 27, 66
Britain
  independent air force 12
  wartime policies 23
British Aerospace 180
  Bae 146: 56
  Sky Flash (Rb 71): 48, *159*, 162, 179
Bromma Airport 29, *90*, 184
Bücker, Carl Clemens 11, 12
Bücker aircraft
  Bü 181 (Sk 25): 30
  trainers 11, 93
Burnett, James A 19

Caproni Ca. 313: 24
Cirrus-Hermes engine 13, *14*
Comair 171, *174*
Congo, Swedish aid to UN 112-14
Convair CV 240: 89
CR 42: 23, 24
Crossair *57,* 58, *168,* 175, *176*
CSF Company 42

Dahlén, Gustav 9
Dahlström, Erik 146
Daimler-Benz Company 30
Daimler-Benz engines
  DB 601: 30
  DB 605B: 27, *28,* 30, 31, *71*, 71, 75, 81
Danish Air Force *see* Denmark
Danish Brigade 62
Dassault Mirage F-1: 54
DDL Company 90
de Havilland aircraft
  Mosquito (J30) 116
  Tiger Moth (Sk 11) 14, *16*
  Vampire (J28): 35, 80, 137
de Havilland engines
  Ghost 37, *38,* 104, 105, *111,* 111
  Gipsy Major X: 93, *94,* 94
  Gipsy Six 13, *15*
  Goblin 32, 35, 81, *82,* 84, 104
De Schelde Company *34,* 96
Defence Committee 54
Defence Department 50
Defence Material Administration (FMV) 50, 163, 180
Dellner, Gunnar *18*
Denmark
  aircraft orders 44-5, *133,* 135, *167,* 167
  NATO procurement *52,* 54
DNL Company 90
Douglas DC aircraft 89, 90, 92, 145
Douhet, Gen Guilio 28
Draken *see* Saab 35 Draken *under* Saab aircraft

E 4 aircraft 14
Edgerton Graier & Germeshausen 130
Electrolux Group 18
Enderlein, Maj O 184
Enoch Thulins Aeroplanfabrik, AB (AETA) 9-10
Enskilda Bank 16, 17
Ericsson Company, LM (Ericsson Radar Electronics) 42, 50, 54-5, 123, 129, 150, 158, 178
Ethiopia
  aid to UN 112
  aircraft sales 62-3, *63,* 94, 96, 97, *102*
  relief operations *166*

F 1 Wing 68, 72, 135
F 2 Wing 62
F 3 Wing 62, 71, 130
F 4 Wing 62, *110, 112,* 115, 130, 162
F 5 Wing (Flying School) 141
F 6 Wing 62, 80, 109, 120, 152
F 7 Wing 62, 72, 80, 83, 109, 120, 152
F 8 Wing 80
F 9 Wing 80, 186
F 10 Wing *81,* 83, 130, 135
F 11 Wing 68, 71, 120
F 12 Wing 62, 80, 124, 135
F 13 Wing 106, *126-7,* 129, 130, 135, *161,* 162, *182*
F 14 Wing 72, 120
F 15 Wing 80, 115, 120, 152

F 16 Wing 129, 135, 162
F 17 Wing 72, 84, 120, 135, 162
F 18 Wing 129
F 21 Wing 71, 130
Fairchild Industries Inc 56-8, 168-73 *passim*
Fairchild aircraft 168-73
Fairey Firefly 63
Falcon missile 45, *46*, 130, *151*
Falk, Maj Gen Greger *51*
Farman biplane 9
Faxén, Torsten 27
Federal Aviation Administration, US 58
Fernberg, Karl-Erik 138
FFVS *see* Flygförvaltningens Flygverkstad i Stockholm
FFVS J 22: 28-30, 182-6
Fiat Company 90
Fiat aircraft
  CR.42: 23, 24
  G.50: 22
Finland
  aircraft orders (Draken) 45, 53-4, *134*, 135 (Safir) 96, 97, *98* (other) 53-4, 63
  missile orders 46
  Swedish aid in war 22, 23
Florman, Carl and Adrian 12, 86
Flygförvaltningens Flygverkstad i Stockholm (FFVS) Company 30, 54
Flygförvaltningens Flygverkstad i Stockholm (FFVS) aircraft, J22: 28-30, 182-6
Flygindustri, AB (AFI) (Malmö Flygindustri) 11, 12, 13, 52, 163
Flygindustri aircraft 12, 163
Flygkompaniets Verkstäder Malmen (FVM) 10-11
Flygkompaniets Verkstäder Malmen aircraft 11
Flygplanverken, AB (AFV) 20-1, 66
Flygplanverken aircraft, GP 9: 20-1
Flygvapnet *see* Air Force, Swedish
Flying School (F5) 141
Focke-Wulf Company 18
Focke-Wulf Fw 44J Stieglitz *18*, 18
Fokker Company 39, 90, 96
Fokker aircraft
  C.VE (S6): *10*, 13
  F.27 Friendship 90
Fontoura, Olavo 90
Förenade Flygverkstäder, AB (AFF) 18, 19, 24, 28
Förenade Flygverkstäder (AFF) aircraft
  AFF/Gassner 19-20
  P7: *22*
Forsberg, Uno 25
Försvarets Robotvapenbyrå 40
Fouga Magister 112
France, Swedish aviation interests 9, 20, 42, 46, 49, 130
Friis, Gen Torsten 14, 16

Gassner, Alfred 19, 20, 66
General Dynamics F-111/YF-16: 54
General Electric engines
  CT7: 58, 169, 174

F404: *55*
J85: 49, *137*, 142
RM 12: *178*, 179
Germany
  aviation interests 11, 12, 23, 104
  licensing 9, 18-19, 20, 30
  missile landing 40
Glen L Martin Company 92
Gloster Gladiator 22
Göring, Reichsmarschall Hermann 30
Götaverken (GV) 15, 20, 66
Götaverken (GV) aircraft, B4/GP8: 20
GP 8/9 aircraft 20-1
Gripen *see* Saab 39 Gripen *under* Saab aircraft
Gullstrand, Dr Tore 42, *52*, 52
Gustafsson, Sten 53

Hagermark, Olle 88
Hägglund & Söner 30, 60, 185
Hägglund & Söner aircraft, Sk 25: 30
Haile Selassie, Emperor 62
Hamre, Gen Sverre *52*, 54
Hansa S5: *11*, 11, 13
Hansa-Brandenburg monoplane 11
Hansson, Per Albin 12, 16
Härdmark, Ragnar 82, *83*, 137
Hawker aircraft
  Hart (B4) *13*, 14, *16*, 20, 22, 27
  Tempest 78
Heinkel Company 11
Heinkel aircraft
  HD 19 (J4): 12
  He 5 (S5): *11*, 11, 13
Hesselmann Company 30
Hoffström, Bo 87
Holm, Tryggve *37*, 37, 39, *41*, *42*, *51*, 51
Honeywell Company 151
Hughes missiles 45, *46*, 130, *151*

India
  aid to UN 112, 113
  interest to Viggen 54
Industrial Commission (IK; 1939) 24-5, 26, 28
Italy
  independent air force 12
  Swedish aviation interests 20, 24, 61
Ivarsson, Tommy *58*, 181

J 1 Phoenix 122: 11
J 4 (HD 19) 12
J 6B Jaktfalken 11, *12*, 14
J 9 (EP-1) 24, 28
J 11 (CR 42) 23, 24
J 19 (L-12) 20, 21, *22*
J 20 (Re 2000) 24
J 21 *see* Saab 21 *under* Saab aircraft
J 22: 182-6, 28-30
J 26 (Mustang) 75, 78-9, 186
J 28 (Vampire) 35, 80, 137
J 29 *see* Saab 29 *under* Saab aircraft
J 30 (Mosquito) 116
J 32 *see* Saab 32 Lansen *under* Saab aircraft
J 35 *see* Saab 35 Draken *under* Saab aircraft
JA 37 *see* Saab 37 Viggen *under* Saab aircraft

Japan, use of Safir 97
JAS Industry Group 54-5, 180
Johnson Group 15
Joint European Airworthiness Group (JAR) 58, 171
Jönköping factory 42, 45
Junders Flugzeug und Motorenwerke AG 11, 12, 19
Junkers aircraft
  F 13: 12
  G 24: 12
  Ju 52/3m: 12
  Ju 86K (B3): 12, 19, 22, *23*, *24*, 68
  K series 12

K 24/37/47 aircraft 12
Karnsund, Georg 57
Katangan Air Force 112, 114
Kazmar, Ernest W 19
Kjellson, Henry 11
Koch, Peter 11
Kockums shipyard 15

L-10 *see* Saab 17 *under* Saab aircraft
L-11 *see* Saab 18 *under* Saab aircraft
L-12 (J19) 20, 21, *22*
L-13 *see* Saab 21 *under* Saab aircraft
L-21 *see* Saab 21 *under* Saab aircraft
Lalander, Kurt *42*
Lampell, Col Sven 112
Lansen *see* Saab 32 Lansen *under* Saab aircraft
Le Rhône engines 9, *10*
Lidbro, N 37
Lidmalm, Tord *42*, 87, *91*
Lignell, Karl 87
Lindqvist, Maj Gen Gunnar 133
Linköping
  company headquarters *53*, 66
  works 16, *17*, 18, *36*, *41*, 45, 62
Lockheed F-104 Starfighter 54
Löfkvist, Hans Erik *42*
Lufthansa Airline 96
Lundberg, Bo 14, 20, 28, 66, 182
Lycoming engines *94*, 95, 96, *99*, 165

McDonnell Douglas Company 55-6
Malmer, Dr Ivar 10, 11
Malmö Flygindustri (MFI) 52, 163 *see also* Flygindustri AB
Malmslätt maintenance works 10-11, 12, 141
Martin 2-0-2: 89
Maverick (Rb 75) missile *148-9*, 150, 179
Maybach engines 11
Mercedes engines 9
MFI-15/17 Safari Supporter 163-7
Mileikowsky, Dr Curt 51, *52*, 53
Moore, Sqn Ldr Robert 105, *106*, 106
Mustang (J26) 75, 78-8, 186

NA-16-45M (SK 14): *19*, 19, *74*, 76
NATO procurement *52*, 54
Navy, Swedish, aircraft and missiles 40, 46, *48*, 59
Netherlands
  NATO prcurement *52*, 54
  Swedish contracts 10, 20, *34*, 39, 90, 96

Nicolin, Curt 40
NK Company 185
Nohab Flygmotorfabriker AB 11, 15-16, 18, 20, 26
Nohab engines
  Mercury 11, 16, 26, 59 *see also* Svenska Flygmotor engines
Nord missiles, CT-20: 46
Nordiska Aviatik AB (NAB) 9
Nordquist, Maj Elis *13*, 13
North American aircraft
  F-86 Sabre 109
  NA-16-4M (Sk14) *19*, 19, *74*, 76
  P-51 Mustang 75, 78-9, 186
Northrop Corporation 19
Northrop aircraft
  8A (B5): 19, *21*, 21, 22, 27
  YF-17: 54
Norving Norway *174*
Norway
  aircraft orders 96, *103*, *167*, 167
  NATO procurement *52*, 54
Nothin, Torsten 20
Nydqvist & Holm (Nohab) 11

Ö1/Ö4 aircraft 11
Ö 9, ASJA 13, *15*
Olow, Bengt *42*, 117, 125, 128
OMERA/Segid Company 130
Ostermans Aero AB 63

P 7: *22*
Pakistan, aircraft pruchase 52, *164*, 167
Pellebergs, Per 162, 170
PERT Planning System 146
Petersson, Tage *58*
Philips Company 42
Phoenix aircraft 11
Piaggio engines
  P VII RC-16: *19*, 24
  P XIbis RC 4: 26, 61, *64*
Piasecki XH-16 helo 92
Piper L-21B: 163
Porat, Gösta von 11
Pratt & Whitney engines
  JT8D (RM8): 50-1, 54, 144-5, *155*, *161*, 161
  R-2000/2180: 87, *88*, 89, 92
  R-4360 Wasp Major 92
  Twin Wasp (supply) 24, 26, 31, 66, 183 (Saab 17/18) 25, 27, *27*, 59, *60*, 61, 66, 68, 71 (Scandia) 86-7 (J22) 28
Project Directorate 145

Raab-Katzenstein RK 26 Tigerschwalbe 13, *15*
Rb missiles *see* Falcon; Maverick; Saab missiles; Sidewinder; Skyflash
Reggiane aircraft, Re 2000 (J20): 24
Republic Express (Business Express) 175
RF-35 *see* Saab 35 Draken *under* Saab aircraft
Rolls-Royce engines
  Avon Series 39-40, 50, 117, 120-1, *126-7*, 128
  Eagle 11

Rosen, Col Carl Gustav von 62, 94-5
Royal Institute of Technology (KTH) 76
Royal Norwegian Air Force *see* Norway

S 5 (He5): *11*, 11, 13
S 6 (Fokker C.VE) *10*, 13
S 16 (Ca.313): 24
S 17BS *see* Saab 17 *under* Saab aircraft
S 18/S 23 aircraft 11
S 18A *see* Saab 18 *under* Saab aircraft
S 29 *see* Saab 29 *under* Saab aircraft
S 32C *see* Saab 32 Lansen *under* Saab aircraft
S/SF/SH 37 *see* Saab 37 Viggen *under* Saab aircraft
SAAB Aktiebolag *see* Saab Company
Saab Company
  foundations 18, 24
  expansion 31, 35-9, 58, 87
  restructuring 20, 50, 51-3, 66, 168-9
  car manufacture 35, *37*, 58, 93
  see also Svenska Aeroplan Aktiebolaget AB (SAAB)
Saab aircraft
  B3La 54, *55*
  B 3: 19, 22, *23*, *24*, 68
  J 19 (L-12): 20, 21, *22*
  MD 80 series 56
  MFI-15/17 Safari/Supporter 163-7
  Saab 17 (L-10) *22*, 25
    (development) 20, 27, 28, 59-62
    (service) 26, 62-3 (details) 65
  Saab 18 (L-11) *27*, *29*
    (development) 26-7, 28, 66-8, 71
    (service) 68-73 (details) 74
  Saab 21 (L-13): *31*, *32*, *37*
    (development) 30, 31, 32-3, 75-8, 81-3 (service) 37, 78-80, 83-4 (details) 28, 80, 85
  Saab 24: 28
  Saab 29 (development) 33, 36-9, 104-11, (service) *36*, 111-15 (details) 115
  Saab 32 Lansen *39*, *40*
    (development) 28, 39, 42, 116-21
    (service) 41, 49, 121-4 (details) 124
  Saab 35 Draken *42*, *43*, *44*
    (development) *41*, 43-4, 49, 125-34
    (service) 44-5, *46*, *47*, 53-4, *121*, 133-6 (details) 136
  Saab 37 Viggen *50*, *53*
    (development) 49-51, *52*, 54, *55*, 143-54, *55*, 143-54, 179 (service) *140*, 148-61, 177 (details) *140*, 161-2
  Saab 39 Gripen 55, 177-81
  Saab 90 Scandia *35*
    (development) 35, *36*, 37, 86-90
    (service) 90-2 (details) 92
  Saab 91 Safir *34*
    (development) 35, *36*, 93-4
    (service) 95-8 (details) 98-103
  Saab 105
    (development) 48, *49*, 137-41
    (service) 53-4, 141-2 (details) 142
  Saab 201/2: 97, *106*

Saab 210: *125*, 125, *129*
Saab 340: *56*, *57*, 57-8, 168-71
Saab equipment
  autopilot 123, 129
  bombsight 27-8, 61, *62*, 71, 80, *119*
  computers 50, 80 *119*, 150
  ejector seat 28, 30, *73*, 76, *78*, 105, *121*, 128, *138*
  jig-making 38
  sighting systems 42, 129, 130
Saab Missiles AB 46-7
Saab missiles
  development 40-1, 45-8, 72, 130
  fitting *118*, *119*, *128*, 131, 133, 148, *150*, 150, *151*
Saab-Bofors Missile Corporation (SMBC) 48
Saab-Scania Aktiebolag 51-3, 168, *see also* Saab Company
SABENA Airlines 96
Safari/Supporter (MFI-15/17) 163-7
Safir *see* Saab 91 Safir *under* Saab aircraft
Salair 58, 173
Salwén, Lt B J E 185
Sandvik Steel Works 9
Scandia *see* Saab 90 Scandia *under* Saab aircraft
Scandinavian Airlines System (SAS) 12, *89*, 90
Scania-Vabis 51
Schröder, Harald *58*, 146
See Fabriks AB 185
Segerqvist, Lennart *13*
Seversky-Republic aircraft
  2P-A (B6): 24
  EP-1 (J9): 24, 28
SF 340 *see* Saab 340 *under* Saab aircraft
SFA *see* Svenska Flygmotor
SH/SF 37 *see* Saab 37 Viggen *under* Saab aircraft
Sidewinder missile
  AIM 9L (Rb74) 47, *159*, 162, 179
  Rb24: *104*, 112, 121, 129, 133, 150, *151*
Sierra Leone, aircraft order 167
Sjöberg, Erik 170
Sk 10/12, ASJA 13, 14, *15*, *16*, *18*, 18
Sk 11 (Tiger Moth) 14, *16*
Sk 14 (NA-16): *19*, 19, *74*, 76
Sk 25 (Bü 181): 30
Sk 35 *see* Saab 35 Draken *under* Saab aircraft
Sk 37 *see* Saab 37 Viggen *under* Saab aircraft
Sk 50 *see* Saab 91 *under* Saab aircraft
Sk 60 *see* Saab 105 *under* Saab aircraft
Skandinaviska Enskilda Banken 51
Sköld, Per Edvin 20
Sky Flash missile (Rb71) 48, *159*, 162, 179
Smith, Claes 59, 78, 88
Söderberg, Maj Gen Nils 25, 28, *29*, 30, 39, 182
Södertelge Werkstäders Aviatikavdelning (SW), AB 9
Sparmann, Edmund 14
Sparre, Claes *18*

Spica missile craft *48*
SRA Company 42
STAL Company 32, 39-40
STAL engines
  Dovern II (RM4): 39, 116-17, 124
  Glan: 39, 40, 117
  GT 35: 40
  Skuten 32
Stockholm Technical Museum 10
Sundén, Col Åke *51,* 82, *83*
Suter, Moritz *57*
Svantesson, Östen 111
Svensk Flygtjänst Company 63, 124
Svenska Aero AB 11-12, 13
Svenska Aero aircraft
  J 4: 12
  Jaktfalken 11, *12,* 14
Svenska Aeroplan Aktiebolaget AB (SAAB)
  foundation 18, 20, 51
  see also Saab Company
Svenska Aeroplanfabriken (SAF) 9
Svenska Flygmotor AB (SFA) 26, 31, 39-40, 51, 61
Svenska Flygmotor engines
  Glan 39, 40
  Goblin 32, 35, 82, 84
  Mercury 61
  Mx 31
  R 201: 39
  RM 2B (Ghost) *38,* 111-12, 116
  RM 5/6 (Avon) 120-1, *126-7,* 128, 129
  RM 8: 50-1, 144-5, *155, 161,* 161
  RM 68B: *126-7*
  Twin Wasp 26, *27,* 27, 28, *60,* 180
  see also Nohab engines

Svenska Järnvägsverkstäderna, AB (ASJ) 12, 13, 20
Svenska Järnvägsverkstädernas Aeroplanavdelning, AB (ASJA) 13-14, 16-20, 59, 66
Svenska Järnvägsverkstädernas Aeroplanavdelning (ASJA) aircraft
  B 4/5: *13,* 14, *16,* 19, *21,* 21, 27
  J 6B: 11, *12,* 14
  L-10: 20, *22, see also* Saab 17 *under* Saab aircraft
  Ö 9: 13, *15*
  Sk 10/11: 13, 14, *15, 16*
  Sk 14: *19,* 19, *74,* 76
  Viking 13, *14, 15*
Swedair 124
Swedish Air Lines (ABA) 12, 22, 29, 35, 86-92 *passim*
Swedish Intercontinental Airlines (SILA) 33, 35, 92

T 18 *see* Saab 18 *under* Saab aircraft
Taiwan, Saab 340 order 173
Target Flying Squadron 124
Thörnell, Gen O 24
Thulin, Dr Enoch *8,* 9, 10
Thulin Company (Enoch Thulins Aeroplanfabrik) 9-10
Thulin aircraft *8,* 9, 10
Thunberg, Lieut Gen Lage *51,* 146
Tiger Moth (Sk11) 14, *16*
Trollhättan works 16, 18, 20, *31,* 35, *39,* 62
Tummelisa (Ö1) 11
Tunisia, aircraft orders *102*
Turboméca engines 49, *137,* 137, 141

Uggla, Erland 13
United Nations, Congo operations 112-14
United States of America
  sales restrictions 19, 20, 59, 182
  Swedish orders 23-4, 169, 171
Uno Särnmark Company 185

V1/2 missiles 40
Vampire (J28): 35, 80, 137
VASP airline 90-1, 92
Västerås workshops 12, 71
Vennerström, Ivar 14
Vickers Viscount 91
Viggen *see* Saab 37 Viggen *under* Saab aircraft
Viking aircraft 13, *14,* 15
Vinten Company 130
Volvo Company 26
Volvo Flygmotor 54-5, *155*
Volvo Flygmotor engines *155, 178,* 179
  see also Svenska Flygmotor
Volvo Penta Company 31
Vreedeling, Henk *52*
Vultee P-48 Vanguard/48C: 24, 182

Wahrgren, Ragnar 14, 20, 28, *32,* 37, 86
Wallenberg Jr, Marcus 20, 25, *41*
Wallenberg Group 16
Walter engines, Gemma radial 13, 14
Wänström, Frid 75
Wenner-Gren, Axel 16, *18,* 18, 26
Westerlund, Capt Anders 107
Westland Lysander 19
Wilkenson, Dr Erik 27, *32*
Wingquist, Sven 16
Wright Engines, Whirlwind R-975E: *15,*